This book tells the story of the Swiss Institute for Nuclear Research (SIN). The institute was founded in 1968 and became part of the Paul Scherrer Institute (PSI) in 1988. Its founding occurred at a time when physics was generally considered the key discipline for technological and social development. This step was unusual for a small country like Switzerland and showed courage and foresight. Equally unusual were the accomplishments of SIN, compared with similar institutes in the rest of the world, as well as its influence on Swiss, and partially also on international politics of science.

That this story is now available in a widely understandable form is due to the efforts of some physicists, who took the initiative as long as contemporary witnesses could still be questioned. As is usually the case, official documents always show just an excerpt of what really happened. An intimate portrayal of people who contributed to success requires personal memories. This text relies on both sources. In addition, the events are illustrated with numerous photos.

Andreas Pritzker was born in 1945 in Windisch (Switzerland). After completing his studies in physics at ETH Zurich, he began working as a researcher and consulting engineer in the private sector, and then as researcher for the Swiss Institute for Nuclear Research (SIN). After five years working on the staff of the ETH Board, he became a member of the management of the Paul Scherrer Institute in 1988, first as head of administration and later of logistics. Since 2003 he has been active as a freelance author and publicist. He is married and lives in Küttigen (Switzerland).

In addition to his contributions to anthologies, his novels *Filberts Verhängnis* (1990), *Das Ende der Täuschung* (1993), *Die Anfechtungen des Juan Zinniker* (2007), *Allenthalben Lug und Trug* (2010), as well as his novella *Eingeholte Zeit* (2001) have been published. He was co-editor of the REFUNA AG story and editor of several historical texts.

The Swiss Institute for Nuclear Research SIN

Andreas Pritzker

Originally published in German as *Geschichte des SIN* by munda, Küttigen (Switzerland) in 2013

Translated from the German by Royal Canadian

Front cover photo: Jean-Pierre Blaser

Produced and published by BoD – Books on Demand, Norderstedt (Germany)

ISBN 978-3-7386-0820-5

More books by Andreas Pritzker
are presented on
www.munda.ch

For all who made SIN's success possible

Contents

Acknowledgments

In January 2012 Jean-Pierre Blaser invited me to a discussion meeting. Also present were Werner Joho and Urs Schryber who had taken the initiative to document the remarkable history of the Swiss Institute for Nuclear Research (SIN) while it was still possible to interview contemporary witnesses.

SIN was founded in 1968 as an annex-institute to the ETH Zurich and, together with the Swiss Institute for Reactor Research (EIR), became the Paul Scherrer Institute (PSI) in 1988. Jean-Pierre Blaser was the founder and only director of SIN. After having managed the project for the fusion of EIR and SIN, he became the first director of PSI. From the beginning, Werner Joho and Urs Schryber were involved as key physicists in the design and construction, and later the operation and upgrading, of the proton accelerators of SIN and PSI. Moreover, they developed the concepts of future large-scale facilities based on accelerators. I worked as a researcher at SIN from 1980 until 1983, and then on the staff of the ETH Board, responsible for the relations to the annex institutes. From 1988 until 2002 I managed first the administration, then the logistics at PSI.

The conversation during this first meeting revealed a plethora of information on SIN. This alone made it clear that SIN occupies an important position in the history of Swiss science.

During the year 2012 I gathered the necessary information on the history of SIN. My most important and most systematic source was their annual reports. In the beginning they were edited by Honorary Doctor A. Brunner, who was active as science editor and wrote broadly understood articles for the Neue Zürcher Zeitung, for instance, about CERN and SIN. Simple in their design, these annual reports, nevertheless, communicate all important information regarding SIN.

Concerning the back story, I was able to rely on the documents created by Hermann Wäffler, Kurt Alder as well as Charles P. Enz et al. Jean-Pierre Blaser summarized important activities of SIN in notes and added amendments. Werner Joho and Erich Steiner helped develop this text and made personal experiences and photos available to me. Because Erich Steiner participated many years in the layout of experimental equipment – later he managed the high-current project and finally for PSI the large research facilities as well as the project spallation neutron source – Joho and Steiner represented the two main activities of SIN – accelerators and experimental facilities. Urs Schryber corrected and amended an early version of the manuscript. Ralph Eichler commented on the manuscript and supplied me with documents of the ETH Board's activities regarding the future of EIR and SIN (including the Hayek Study) as well as the B-Meson Factory. Additional documents regarding the Hayek Study were lent to me by Karsten Bugmann. In 2012, Dieter Brombach delivered to the PSI library precious and partially lost documents. Urs Brander, the manager of the PSI library, was very helpful in my search for documents. I received important tips and corrections regarding the entire manuscript from Christoph Tschalär and Wilfred Hirt, and from Eros Pedroni regarding the particle physics therapy.

General information was accessible to me on the Internet. Especially productive were the official publications of the Swiss Confederation, the Historic Lexicon of Switzerland as well as the knowledge portal of the ETH library in Zurich.

Photos were made available to me through the kind cooperation of PSI's photo library, Werner Joho, Erich Steiner and Jean-Pierre Blaser. Christa Markovits also contributed photos. Certain photos were taken from the annual reports of SIN. Some photos I was able to acquire from the ETH library. CERN as well as the Lawrence Berkeley National Laboratory made high-resolution photos available to me. I also found several photos on the Internet.

Retired secretary of state Heinrich Ursprung volunteered to describe his impressions of SIN in a preface – he accompanied the institute for years, as president of the ETH Zurich, and then as president of the ETH Board. Finally, Jean-Pierre Blaser summarized, in a personal retrospection, his overall view of SIN as well as his reflections on the politics of science, in addition to all his other information that became part of the manuscript.

The printing of the original German version of this book was financially supported by ETH Zurich and PSI Villigen.

My heartfelt thanks go to all who contributed to the success of this work.

Andreas Pritzker

Preface by Heinrich Ursprung

The planning and further development of universities are challenging tasks. They demand the consideration of a large time scale. If an existing course of study is to be eliminated, the brake length is measured in semesters; a little less for a Bachelor than a Master, a little more for a PhD. The addition of a new research field into the program demands the development of research course content and the projection of needed infrastructure. Since science policies cannot be carried out without scientists, sooner or later the question of (new) leaders arises.

The planning and further development of physics programs at ETH Zurich has taken place in exemplary fashion. In the 1950s, the step from nuclear physics to particle physics was in the air. CERN was founded on an international level. Advanced thinkers in various countries thought that physicists, who had prepared their experiments at home, would have access to the fantastic facilities and the immense experimental possibilities of CERN. Among Swiss physicists, Paul Scherrer was an early leader regarding the transition from nuclear physics to high-energy physics. As a theoretician, Res Jost recognized the potential of particle physics also in theoretical terms. And the president of the ETH Board, Hans Pallmann, professor of agricultural chemistry, had the strength of purpose to support the physicists' ambitious plans back in 1959. On the strength of his application, the Federal Council elected Professor Jean-Pierre Blaser as director.

*

On October 1, 1973 I took over the presidency of the ETH Zurich. In this position I had the pleasure of meeting with SIN director Blaser on a regular basis, especially in the presidents' committee of the ETH Board. This panel, with its somewhat disparaging name, met under the direction of the president of the ETH Board and consisted of the presidents of both ETHs, Zurich and Lausanne, and the directors of the five research institutes active at that time. Everyone represented the interests of his own institute to the extent that they fell under the responsibilities of the ETH Board. For more than one and a half decades I witnessed the manner in which Blaser developed, managed and represented the creation and further development of SIN. His votes were mostly brief and always logical. As a fast thinker, he gave easily understandable responses to questions during discussions. Not all members of the panel possessed the same confidence based on background as he did; this sometimes led to tensions which never deterred him. The official languages during these sessions, as well as the ETH Board's plenum, were German and French, and Blaser had mastery of both.

I experienced discussions about SIN in a number of other committees, alone or with Blaser, in committees of the Swiss Parliament in preparation for parliamentary decisions, in work groups with Federal Offices in discussions over financial bills, in science policy committees. Often the question was raised whether SIN could fulfill its task as a user laboratory, or whether self-interest guided certain expansion plans. Such questions were usually raised, directly or indirectly, by university circles whose Federal financial grants were in close proximity to those of the ETHs. My visits to CERN, DESY, in Grenoble, Los Alamos, Brookhaven, and Oak Ridge had made it clear to me that good user laboratories are subject to such suspicions. A management that has made excellence of research conducted at its institution a hallmark, has the right to act elitist.

Pretty soon it was clear to me that SIN had developed into a place of research that did not have to fear comparison with other internationally well-known institutes.

*

At the end of 1987 the Federal Council decided to combine both research institutes EIR and SIN into PSI. Now the search was on for a manager. As president of the ETH Board I was being swamped with advice that was all identical: Neither the EIR director nor the SIN director could be considered for the management role; no, only an outsider would be able to fuse "both cultures" into a new whole. I listened to the arguments, but I did not believe them. After all, Blaser knew both "cultures" – to the extent that they existed – by being part of them for years. He had supported the merger and had meticulously prepared for it. He knew how to combine existing strengths and increase their effect. He knew how to motivate staff to peak performance. The Federal Council elected Blaser as director of PSI.

This was a stroke of luck for Swiss science. Blaser handled his responsibilities with bravado and thereby created the foundation that would blend the strengths of SIN and EIR into the charisma of PSI.

Dear Jean-Pierre: My heartfelt thanks!

Heinrich Ursprung

1 The Back Story

1.1 The Introduction of Nuclear Physics at the ETH in Zurich

The 1930s were marked by a turbulent development of the physics of the atomic nucleus. The discovery of the neutron (see glossary of technical terms in the addendum) by James Chadwick in 1932, and the first-time manufacture of radioactive isotopes by Frédéric und Irène Joliot-Curie in 1933 led to a model of the atomic nucleus which was made up of neutrons and protons, and which could be used as a basis for systematic research. Paul Scherrer, who led the Physics Institute of ETH at that time, was fascinated by the new research direction. He recognized its future possibilities early on and decided to introduce it at his institute. This decision was the beginning of an outstanding advancement of nuclear physics in Europe at ETH.

The experiments in nuclear physics all followed the same pattern, that certain materials are being bombarded with fast particles – as for instance, protons or deuterons (heavy water atoms). The mostly positively charged projectiles needed to be accelerated to a sufficiently high energy so that, in spite of their electrostatic repulsion with the also positively charged nuclei, they would collide and create a reaction. The examination of the reactions and their products made it possible to gain new knowledge of nuclear physics.

The research instruments for nuclear physics were, therefore, particle accelerators. Basically they were high-voltage generators. They created a strong electrical field which accelerated charged particles. The unit of measure for the generated particle energy was the electron volt (see glossary).

Over the years ETH built several such facilities. The experiences showed quickly that with the increasing energy of the projectiles more complex nuclear reactions could be achieved, which in turn yielded more information to be analyzed. In the beginning, the accelerators did not yet create high particle energies. Since 1936 the ETH used a so-called cathode ray tube which generated up to 140 kilovolts. In 1938, at the suggestion of Paul Scherrer, who provided the necessary means, Hermann Wäffler built a ribbon generator, according to van de Graaff, on two floors of the Gloriastrasse physics building which he was able to put into service in 1940. This created an acceleration voltage of 800 kilovolts. Wäffler and his colleagues used this device for ten years, especially for measuring the nuclear photo effect.

The Swiss National Exhibition of 1939 was the occasion for the construction of a facility by the Physics Institute of ETH, in cooperation with the Zurich company Micafil, that would create high voltage by using direct-current, the so-called tensator (see fig. 1). It was based on the principle developed by Cockroft and Walton and used a switching mechanism for the multiplication of voltage which had been developed by Swiss physicist Heinrich Greinacher. It was being shown at the National Exhibition and then housed in one of the underground chambers of the physics building which had been used for previous research. The man in charge of this equipment was Werner Zünti. The tensator reached a voltage of about 700 kilovolts and would be in use until the 1960s, especially as a neutron generator.

Fig. 1: Micafil Tensator at ETH

This afore-mentioned equipment served to accelerate particles under constant direct-current. If the particles started at zero, they were being directly accelerated by high voltage and left the acceleration path at the desired energy. With the available technical possibilities for creating high voltage, the physicists of the 1930s reached a temporary upper limit of 3 megavolts.

In the United States, Ernest O. Lawrence, together with his colleague M. Stanley

Livingston, took a new path for the creation of higher energies. He built the first cyclotron (schematic view see below) at the University of California in Berkeley in the early 1930s. With the continuing development of the first unit, he was able to reach energies of 6.3 megavolts for deuterons. This opened new research possibilities worldwide.

With this equipment, Lawrence fabricated many heretofore unknown radioactive isotopes of known elements and sometimes even of completely new elements. In 1941, with an even higher-performance cyclotron, he was able to create mesons, known as part of cosmic rays, for the first time. Later he worked on applications of the cyclotron in medicine and biology.

Fig. 2: M. Stanley Livingston and Ernest O. Lawrence in front of the 27-inch cyclotron in the old radiation laboratory of the University of California, Berkeley, 1934.

Lawrence got his cyclotron idea after reading the work of Rolf Wideröe about a linear accelerator in 1928. He realized that in a homogeneous magnetic field the rotational frequency of the particles is independent of their energy (isochronic). Therefore, with a constant high-voltage frequency, particles which fly through the acceleration path simultaneously can continuously be accelerated regardless of their radial position. This results in a continuous beam and a much higher intensity than with a pulsed beam, as in the subsequently developed synchrotron, for instance.

This new acceleration principle found great interest worldwide. In the late 1930s, several European institutes began building cyclotrons. Paul Scherrer, who followed these developments with keen interest, also wanted to have such an instrument for the ETH.

The main components of a cyclotron, i.e., a large electro-magnet and a powerful high-frequency generator, were expensive. Its construction required professional knowledge which precluded building it in-house. In addition, such equipment could not be accommodated in existing institute facilities because of radiation protection; it required a separate building.

As Hermann Wäffler writes in his story about nuclear physics at the ETH, the conditions for starting such a large project were favorable in the early 1940s. In the meantime, experience had been gathered with the building of accelerators, which increased the chances of finding a solution to any project-related technical problems. The ETH had also eased the conditions of employment at the Institute of Physics, and since the outbreak of the war had made a career in a foreign country impossible for an indeterminate length of time, young physicists were willing to engage for some time to come in building a cyclotron at ETH.

The Cyclotron

The classic cyclotron consists of a large electro-magnet, with a flat round vacuum chamber located between its poles. Inside that chamber are two hollow, semi circle-shaped metal chambers which, because of their D-shape, are called "dee". The acceleration gap is located between the chambers. Affixed to the outer edge of the chamber there is a diversion condenser (septum) which is used to extract the particle stream in the direction of a specific goal.

The ions – protons for instance – are injected from their source into one of the chambers under low energy. The magnetic field guides them into a circular path. Since the ions are accelerated each time they pass through the acceleration gap, the radius of their path within the magnetic field increases. This results in a spiral-shaped path from the center to the outer edge. The electric field in the acceleration gap is created by a high-frequency alternating voltage the size of about 100,000 volts. The frequency of the high-voltage must be in sync with, or a multiple of, the rotational frequency of the ions so that they match the appropriate phase as they pass through the gap and are accelerated (rather than braked).

Therefore, in a classic cyclotron, two acceleration paths are operative per rotation. Because the acceleration paths are used over and over again, circular accelerators (cyclotrons, synchrotrons) are more efficient than linear accelerators and also much more compact.

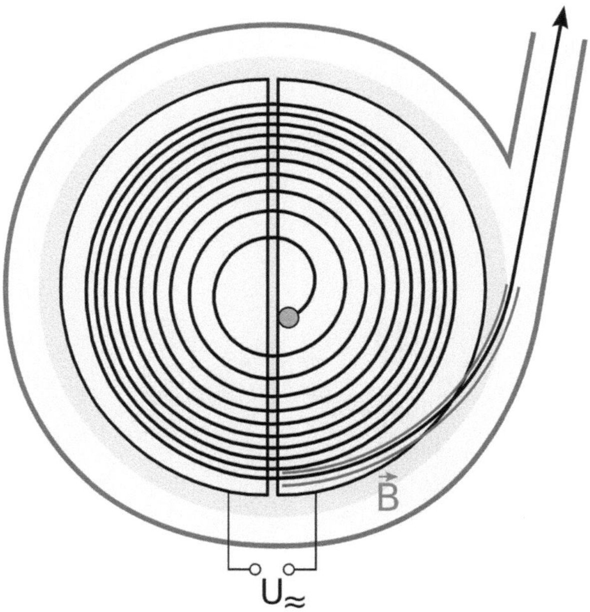

Fig.3: Schematic of a cyclotron (view of the median plane) with decreasing spaces between the particle paths towards the rim, and the extraction of the beam with a bent parallel-plate condenser.

The problem was how to finance the project. Scherrer decided to use a series of lectures in order to win prominent business people as donors. Thanks to Scherrer's highly developed art of making principles of physics easily understandable, his success was overwhelming. The institute received prominent donations for the fund of the cyclotron project. Walter Boveri, delegate of the Brown Boveri & Cie (BBC) board of directors, a friend and benefactor of Scherrer's, energetically supported the project. The company took on the building of the high-frequency generator and also made available qualified personnel for constructing the facility at the institute.

The Maschinenfabrik Oerlikon (MFO) handled the construction and building of the 40-ton electro-magnet. Most of the remaining components could be designed and manufactured at the Institute of Physics and resulted in some new ground being broken. In charge were postgraduates Peter Preiswerk, Pierre Marnier and Jean-Pierre Blaser. The ETH financed the construction of the new building for the cyclotron. It consisted of a square room of 240 square meters, and was 3 meters high on the inside. Because of the necessity for radiation protection, the building was constructed 3 meters deep into the ground and was connected to the institute building by an underground corridor.

Fig. 4: The ETH Cyclotron

The ETH cyclotron (fig. 4) made nuclear reactions with protons and deuterons possible and was projected for energy generation of 15 megavolts. On commissioning the unit, an unexpected effect was observed that was not understood. The rotational direction of the particles should have been symmetric, thereby allowing both directions. However, it was asymmetric and did not match the construction of the unit, so that the beam could not be extracted. At first, only samples could be irradiated inside the chamber. Since this asymmetry could not be understood or corrected – this was the time before computers! – the physicists had to rebuild the unit by turning the cyclotron chamber around. That allowed the particle beam to be extracted.

From 1944 until 1964, the ETH cyclotron was used exclusively for the acceleration of protons. As it turned out, the phenomena generated by the proton reactions gave rise to so many different, interesting questions that they kept the cyclotron work groups fully occupied.

When the exploitability of the ETH cyclotron slowly came to an end, the time had come to think about more powerful equipment for nuclear physics. For that purpose Scherrer formed a cyclotron planning team (see paragraph 2.1).

1.2 *The Step towards High-Energy Physics: The Founding of CERN in Geneva*

The research results of the 1930s and 1940s in the area of nuclear physics encouraged physicists to increase the energy of accelerated particles by an order of magnitude. This trend towards high-energy physics, just like other similar developments, had their start in the USA. It had the result that the particle accelerators got constantly larger and technically more complex. Not only their construction, but also their operation became ever more costly and demanded ever more professionalism. This exceeded the technological capabilities of single university institutes. Therefore, several countries founded national research institutes. Even an international effort became logical for particle physics with highest energy demands and huge, expensive equipment.

When World War II ended, war-torn Europe no longer led the world in science. Therefore, a number of well-known European scientists in the area of atomic and nuclear physics – among them Louis de Broglie, Pierre Auger and Lew Kowarski in France, Edoardo Amaldi in Italy and Niels Bohr in Denmark – took the initiative to found a European Organization for nuclear physics research. A European research laboratory would promote the collaboration of researchers in various countries on the one hand, and on the other hand, the costs of the ever more expensive equipment would be borne jointly.

A UNESCO Conference held in Paris in 1951 became the framework for eleven European countries to sign an agreement for the founding of a "Conseil Européen pour la Recherche Nucléaire" (CERN). The council began its planning work. Already at its third session in 1952 (fig. 5) – which, by the way, was presided by Paul Scherrer – the council determined that Geneva would be the location of the future laboratory. In 1953 the Canton of Geneva agreed to the project through a referendum, and the CERN agreement was concluded the same year. After ratification of the agreement by the now twelve participating European countries, Switzerland among them, the European Organization for Nuclear Research was founded. It retained the name CERN.

Swiss groups participated in CERN from the outset, since Switzerland did not want to miss the connection to high-energy physics. In the beginning, Switzerland contributed about 3 million francs annually to this European research community.

Fig. 5: Third session of the provisional CERN Council on October 4, 1952 in Amsterdam. In addition to deciding on Geneva as its location, the Council decided to build a 25-30 GeV proton synchrotron. Presiding Paul Scherrer can be seen in the back at left and to his right at the same table, Pierre Auger, Niels Bohr, Lew Kowarski and Peter Preiswerk.

In 1957, CERN began accelerating particles with the 600 MeV synchrocyclotron. This allowed experiments not only in nuclear but also in particle physics. After 1964 this equipment was used primarily for nuclear physics since the new proton synchrotron with 28 GeV had been available for particle physics since 1959. Both machines had a long usable life. Later on, between 1967 and 1990, the synchrocyclotron supplied protons for the ISOLDE facility. It made possible the production of many different instable radioactive ions for experiments which encompassed everything from pure nuclear physics up to solid-state physics and medical applications. For a short time, the proton synchrotron was the accelerator with the highest particle energy worldwide. After CERN had built new accelerators in the 1970s, it was used primarily as an injector for new machines.

How did Switzerland end up with CERN?

Jean-Pierre Blaser tells the following anecdote: Pierre Auger had sent a circular, suggesting the founding of CERN, to the directors of all European institutes of physics, Paul Scherrer among them. Scherrer was not in favor of such a project and tossed the circular into the wastepaper basket. Peter Preiswerk, titular professor at the ETH, had participated in a discussion about a European laboratory for high-energy physics at a meeting of the Committee for European Scientific Collaboration back in 1950. He recognized the importance of such an initiative. He rescued the circular from the wastepaper basket and persuaded Scherrer to respond positively. Scherrer finally understood that only a joint European effort had a chance of catching up with the advantage the USA had gained in nuclear physics. He fully supported the realization of CERN and the selection of Geneva as its location. Preiswerk would later also contribute significantly to the development of CERN. In 1954 he took on the leadership of the CERN building department and in that function was responsible for the design and construction of the buildings and installations in Meyrin. Finally, from 1961 until 1971, he became the manager of CERN's Department of Nuclear Research.

2 The Meson Factory: From Vision to Decision

2.1 *The Cyclotron Planning Team*

When the usable life of the ETH cyclotron was slowly coming to an end (see paragraph 1.1), Paul Scherrer, together with the Institutes of Physics of the Universities of Basel and Zurich, assembled the so-called Cyclotron Planning Team in the second half of the 1950s. The team was planning the building of a new, larger accelerator with a budget of 10 million francs. Although the planned machine was intended for nuclear physics, and not for high-energy physics, its costs to build and operate represented a new order of magnitude, as compared with the existing ETH cyclotron. That was also the reason why ETH tackled the project in cooperation with the Universities of Basel and Zurich.

The team was led by Peter Stähelin. As a former student of Scherrer's – he would later function as research director of the Deutsche Elektronen-Synchrotron (DESY) from 1960 until 1967 – he had brought with him the concept of a cyclotron for heavy ions on his return from the USA. The planned concept was equivalent to the 88-inch cyclotron located at the University of California in Berkeley.

Fig. 6: From left – Bob Smith, Hans Willax and Elmer Kelly during the testing of the 88-inch cyclotron on December 12, 1961.

A few members of the team, i.e., Hans Willax, Hans Baumann and Paul Lanz, were sent there for training by Stähelin. A photo from the web page of the Lawrence Berkeley Lab (fig. 6) shows Hans Willax with the American colleagues during the testing of the cyclotron in December 1961.

In the early 1960s, the continuation of dynamic development of physics and its international importance was highly desirable for the ETH Board as the supervising agency of ETH, and especially for its then-president, Hans Pallmann. Also, ETH felt destined to seize initiatives of international importance.

This fit nicely into the political environment of the time which considered the large scientific and technical developments during and after the war as relevant to society. Especially physics was considered central to science. Nuclear physics and solid-state physics advanced to major status. They had created important discoveries, such as nuclear energy, semiconductor electronics, laser and maser, radar as well as imaging applications in medicine.

The ETH Board devoted a number of sessions during 1958 and 1959 to the development of physics at ETH. Subjects were, independent of the impending resignation of Paul Scherrer, the establishment of two new professorships, the work on the planned cyclotron as well as the succession of the late theoretician Wolfgang Pauli who had died in 1958. In its session of January 1959 it was mentioned that the ETH Professors Scherrer, Marmier, Busch and Jost, as well as Professors Staub and Heitler of the University of Zurich and Professor Paul of the University of Bonn, recommended a quick election of Jean-Pierre Blaser, director of the Neuenburg Observatory and docent at the local university, as professor at ETH. His candidacy had been considered previously by the ETH Board, but was then postponed "so as not to rob the cantons' teaching and research institutes of their best people without necessity." But now the Board considered Blaser's qualifications so important to the task that their previous consideration was set aside. The minutes of the session stated: "He is one of few Swiss physicists who knows high-energy research from his own work and has published on the subject of meson physics....He is versatile, experimental and a very talented educator." The ETH Board decided to apply to the Federal Council for the election of Blaser with a start date of October 1, 1959.

Upon the assumption of his office, Jean-Pierre Blaser took over the cyclotron planning team from Scherrer. In consideration of Scherrer's retirement in 1960, he began to newly evaluate the goals of current projects at the Physics Institute of ETH. Influenced by world-wide developments, Blaser's goal was a machine that would make participation in high-energy physics possible, rather than the cyclotron supported by Stähelin.

Such participation was highly topical. In June 1959 there had been a conversation between the president of the ETH Board, Pallman and Res Jost, professor of theoretical physics at ETH, on the subject of high-energy physics. Jost considered high-energy physics a fertile field of activity, with the possibility of gaining deeper insights into physics. In consideration of soon being able to use CERN's proton synchrotron, whose experiments were already in the planning stages, he recommended to form a high-energy physics group without delay. The discussion included the possibility of a national machine of typically 600 MeV. According to general opinion, only countries with larger nuclear physics facilities would be able to conduct efficient research at CERN.

Jean-Pierre Blaser

Jean-Pierre Blaser grew up in Zurich after his father, director of the commercial college in La Chaux-de-Fonds, was called to serve at the high school in Zurich. Nonetheless, Blaser continued to keep up with his connections at Neuchâtel. He concluded his studies at the state college in Zurich and began his studies of chemistry and mathematics at the University of Zurich. At the recommendation of Professor Louis Kollros, who taught geometry at ETH and who was, like Blaser, from La Chaux-de-Fonds, Blaser switched over to ETH in order to study physics. However, he could not find an appropriate diploma thesis with Paul Scherrer and decided to write his thesis under Professor Georg Busch on infrared spectroscopy in 1948. Blaser was interested in astronomy but could find no common ground with that professor and, therefore, became an assistant to Paul Scherrer, together with Pierre Marmier and Peter Preiswerk. With these colleagues he contributed significantly to the building of the ETH cyclotron. He concluded his studies in 1952 with a dissertation on proton-neutron reactions. In between, he spent one year at the University Sanatorium in Leysin because he had become infected with tuberculosis during his military service.

From 1952 until 1955 he spent time at the Carnegie Institute of Technology in Pittsburgh, where he did research in particle physics on the 440 MeV synchrocyclotron.

In 1955 he was named director of the Observatory of Neuchâtel, and in 1956 he became professor of astrophysics at the University of Neuchâtel. His research fields included astronomy as well as the new standard time based on the atomic clock.

In 1959 the Federal Council named him Professor of Physics at the ETH Zurich. From 1962 until 1970 Blaser managed the ETH laboratory for high-energy physics with experiments at CERN.

In 1968 he founded SIN and was its first and only director until 1987. Beginning in 1986 he managed the ETH Board's project of the merger of EIR and SIN and in 1988 became the first director of PSI, a position he held until his retirement in 1990.

His quick intellectual grasp and his willingness to examine statements critically are considered to be the most remarkable traits of Jean-Pierre Blaser. He would immediately ponder assertions of conversation partners if they did not make sense to him and was, thereby, able to assess them realistically. If he found no basis for such assertions, he would respond to them accordingly.

2.2 *The Idea of a Meson Factory*

Considering his move towards high-energy physics, Jean-Pierre Blaser also had visions of building a national center at which all universities could conduct their teaching research. It was clear that the means for such a project would have to come mostly from the Confederation. Therefore, Blaser suggested that such a national research laboratory – the future "Swiss Institute for Nuclear Research (SIN)" – would be built and operated by ETH, but would be made available to all universities. This, by the way, was analogous to the organizational structure of the large nuclear research institute in Los Alamos, New Mexico, which was being run by the University of California.

The next step was to decide on an appropriate acceleration principle. In the 1950s, several accelerators with energies of approximate 500 MeV had been built worldwide. They made it possible to artificially create the π and μ-mesons, which had been discovered in cosmic radiation, in large numbers with the help of the new synchrocyclotron principle in which packets of protons are being accelerated "in sync" under consideration of their increasing relative mass through speed. Important discoveries in physics could be made with such machines at that time, for instance, at the Argonne National Laboratory near Chicago, at the Joint Institute for Nuclear Research in Dubna near Moscow and last, but not least, at CERN. This created the need for increasing the particle currents, which were limited with the synchrocyclotron, by a factor of 100, with the goal of not just studying the mesons, but to also use them as tools. This meant building so-called meson factories.

This goal was being studied intensively in the USA, in Canada as well as in Europe. In Europe, the "European Committee on Future Accelerators (ECFA)", which was close to CERN, recommended the realization of such a facility, in parallel with the development of accelerators with considerably higher energies but only low particle currents. For this purpose, extensive studies were initiated in Germany and also at ETH. Especially the French acceleration specialist Pierre M. Lapostolle of CERN was a strong supporter of such a meson factory in Switzerland. Such a facility was intended to replace the 600 MeV proton synchrocyclotron at CERN.

The primary goal was the acceleration of protons. There were several possibilities of designing the accelerator, but the most important problem was how to control the radioactivity. With the desired beams in the neighborhood of 100 microamperes and up to 1 milliampere, the most difficult problem was the complete deflection of the beam from the accelerator. Protons, which had separated from the beam, slammed into the walls and created radioactivity there. The types of losses that had heretofore been unavoidable with cyclotrons would cause hardly controllable radioactivity.

Interestingly, the development would occur in three very different directions. Because sufficient resources were available, the research center in Los Alamos (USA) chose a linear accelerator. It allows the accelerated beam to exit the forward end without problems, but at a length of almost one kilometer, it is huge and uses an enormous amount of energy.

Fig. 7: LAMPF Accelerator

A highly interesting physics principle was chosen by the TRIUMF Research Center in Vancouver (Canada): A cyclotron which accelerates negative hydrogen ions instead of protons. When those ions are being directed through a foil at the end, which strips them of their electrons, they turn into protons and exit the magnetic field of the accelerator unaided. However, one problem is that such "stripping" is likely to also occur through the magnetic field of the accelerator which leads to losses as well as radioactivity. This calls for a weak magnetic field and, therefore, extremely large and accordingly expensive magnets. In addition, a very tight vacuum is necessary.

Fig. 8: TRIUMF Accelerator

LAMPF and TRIUMF

LAMPF (Los Alamos Meson Physics Facility) is part of the Los Alamos National Laboratory in New Mexico (US), a research facility of the American Government. It is operated by the University of California for the Department of Energy (formerly the Atomic Energy Commission). Los Alamos is known for having developed the first atomic bombs under the name "Manhattan Project". In 1972 the most powerful linear accelerator at that time was being commissioned in Los Alamos. It was mostly nuclear research that was conducted with protons that had been accelerated to 800 MeV. Today, this accelerator is used for operating a spallation neutron source.

TRIUMF (Tri-University Meson Facility) was founded in 1968 by three universities in British Columbia (Canada) in order to fill research needs that were beyond the means of a single university. It houses the world's largest cyclotron with a magnet diameter of 18 meters and a weight of 4,000 tons. Protons are accelerated to 500 MeV. Originally, TRIUMF made possible research in nuclear and particle physics. Later on, molecular biology and materials science as well as nuclear medicine were added. The development of new accelerators also belongs to the core business of TRIUMF.

LAMPF and TRIUMF, together with SIN (today PSI) are the only three meson factories in the world. The facility at SIN (today PSI) is the one with the most intensive proton beam.

2.3 *The Concept of Hans Willax*

The cyclotron planning team of the ETH chose a "normal" cyclotron with which protons were to be accelerated directly, whereby the high energy made a so-called sector focusing necessary. Since the physicists were primarily interested in meson research, rather than accelerator development, they investigated the purchase of such a machine. Indeed, many companies were active in building accelerator facilities at that time. Several quotations were investigated. The most auspicious one at that time seemed to be an offer from the German AEG for a 450 MeV isochronal cyclotron, using a continuous beam, rather than a pulsed beam as in a synchrocyclotron, of 100 microamperes of current. After an in-depth examination by the ETH planning team, now led by Willax, there were doubts whether 100 microamperes could actually be reached, and also the energy was at the lower end for creating mesons.

All of these considerations led Willax to once again deal with the basic questions in 1962. He was convinced that a so-called sector-focused isochronal cyclotron represented the best basis for a compact, economical accelerator in a meson factory, as long as it would be possible to safeguard the almost complete extraction of the beam. It was clear to him that for this purpose the proton paths up to full energy needed to remain spatially separated. All this required high-frequency voltages far above those that had ever been achieved with a cyclotron. Achieving the desired energy of 500 to 600 MeV necessary for producing mesons also made new demands on the structure of the magnet in order to assure sufficiently strong focus under relative increase of mass of 160% of rest mass.

Willax came up with the whole new idea of sectioning the magnet off into eight separate, spiral-shaped sectors and to embed between them four high-frequency cavities (fig. 9). With this structure – especially with the spatial separation between magnetic field and high-frequency system – he achieved all at once the necessary focus, and the cavities supplied the necessary extremely high acceleration per revolution with high effectiveness, thus creating separate particle paths.

However, the price to be paid for this ingenious idea was the drawback that such an accelerator could not begin at energy zero. It needed a so-called injector cyclotron which would accelerate the protons to about 10% of their final speed. Therefore, the center was missing from the main accelerator, which is why it was and still is named "ring cyclotron" or "ring" for short.

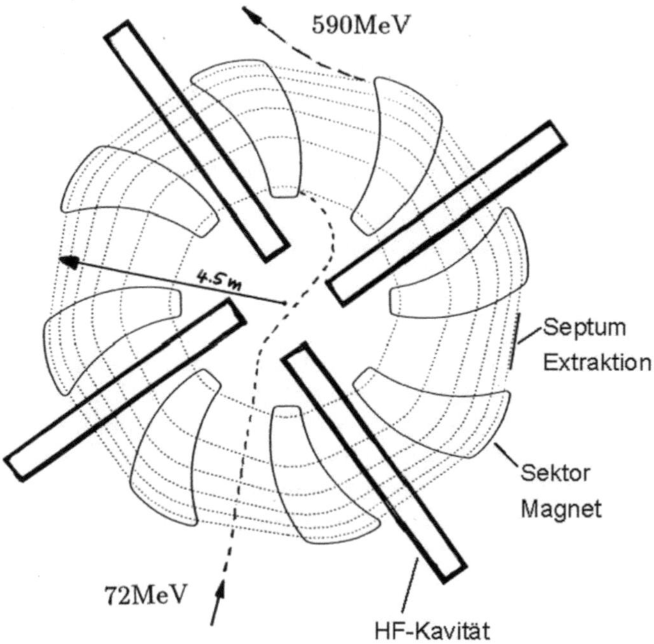

Fig. 9: The Willax Concept

Hans Willax

Hans Willax was born in 1929. He concluded his physics studies at the Technical University in Munich with a PhD in nuclear physics. In 1959 he joined the ETH cyclotron planning team. In order to work on a project of a spiral cyclotron for nuclear research, ETH sent him to the University of California, Berkeley, for two years. After his return in 1962, he became an essential contributor to the "meson factory" project, a challenging task that would become his life's work. He suggested a two-stage acceleration concept which promised high proton currents at high energy and, therefore, intensive pion rays. Willax realized his concept during the following years at the newly founded SIN. For the internationally recognized acceleration specialist it became the zenith of his career when in 1974, using his concept, protons were accelerated to 600 MeV and extracted from the accelerator for the first time.
During the development and building phase, the unusual capabilities of Hans Willax came into effect: In-depth knowledge, innovative ideas and the talent to turn a large project into

reality from an engineering standpoint. That he was able to combine these talents with great human faculties of perception for the concerns of his colleagues, would become another key to his success.

After Hans Willax had succeeded in bringing his ring cyclotron gradually up to the specified capacity, he turned his attention to a new challenge. In 1980 SIN sent him to the Nuclear Research Facility in Jülich, where he participated in the German study of spallation neutron source. While doing this work, he contracted an illness to which he would succumb in April 1981.

The two-stage Willax concept caused great enthusiasm among the members of the ETH team. When Jean-Pierre Blaser presented the new direction of the project to the president of the ETH Board, Pallmann, it caused Pallman concerns. He feared that the proprietary development would cause delays and uncertainties. And he was right. Since a meson factory had become the new goal, members of the original Basel/Zurich/ETH group had been admitted to the cyclotron planning team, if only to emphasize the national aspect. There was resistance among those members, and all of a sudden expert opinions appeared, saying that the Willax concept could not work. Since these opinions were issued from the recognized stronghold, Berkeley, Willax felt rightfully that he had been backstabbed, and great tensions arose among the planning team. To defuse the situation, the Aargau Councilor of State, Robert Reimann, was asked to act as mediator.

Reimann was a member of the National Research Council and, therefore, familiar with the universities. A great many technical and political discussions as well as committee sessions followed, leading to a vote of confidence for the acceleration concept of Willax in 1963, making it possible to continue the meson factory project on the basis of his concept. During the same year, the concept was presented to the experts during the 3rd International Cyclotron Conference at CERN where it gained much respect.

The first acceleration stage – the injector cyclotron – was to be purchased from a commercial source. In consideration of the funding proposal, with which the Federal Council would apply for the meson factory financing to the Federal Parliament, project members began negotiations with several firms. Offers were quickly received – such as from AEG and Philips – however, new ideas came up regarding the use of the injector. Based on specifications, the accelerator builders favored the machine from AEG. However, the Physics Institutes at Basel and Zurich wanted to use the injector themselves for the purpose of nuclear research, and for that purpose the machine from Philips was better suited. At first blush, it did not seem to make sense to shut down the main accelerator part-time so that nuclear experiments could be conducted with the injector. Nevertheless, the project leaders decided on the cyclotron from Philips. Incidentally, this was in accordance with a modified form of machine for nuclear physics which the cyclotron planning team had conceived originally.

Although at the time, years prior to commissioning the facility, no one dared bring up the subject, Willax and Blaser were having their own deliberations. They were convinced that the ring cyclotron would be able to create beams up to 1 milliampere, so that perhaps relatively soon the need for a new injector would arise and that such a machine would have to be built in-house and be based on the ring cyclotron principle. An accelerator made purely for the purpose of injection, such as the AEG machine, would then have to be written off. The multi-use cyclotron from Philips, on the other hand, could continue to be of use. And since nuclear physicists with their teaching-related research would be making valuable contributions to the nationally desired institutional diversification, this idea was certainly logical. It turned out to be advantageous not only from the standpoint of science policy, but also would later allow the development of a first medicine project at SIN – the treatment of melanoma of the eye – under favorable circumstances.

Accelerator Development 1962

The following episode illustrates the conditions of accelerator development at ETH at the time. In January 1962, Werner Joho approached Professor Jean-Pierre Blaser regarding a subject for his thesis. Blaser suggested that he calculate the particle beams in a cyclotron with a computer. For Joho this meant first spending three months at the computer center in Uppsala, where he learned the FORTRAN computer language, and then a deployment to CERN in Geneva where the only available super-computer in Switzerland was located. After finishing his thesis, Joho was accepted into the cyclotron planning team by Blaser and was named group leader of beam dynamics.

2.4 The ETH Funding Proposals 1965 and 1972

The next step was planning the funding. ETH Board president Pallmann decided to integrate it into the ETH funding proposal for buildings of the Federal Council for the year 1965 to be submitted to the Parliament. This building budget was decisive for the development of ETH because it included not only the meson factory in Villigen, but among other things also the new campus – the so-called ETH satellite – in the Höngg suburb of Zurich to be dedicated first to physics. This budget meant debt to the tune of 444 million francs. Of those, 218 million were planned for the Höngg campus, and for the "construction of a research facility for nuclear physics with an accelerator of high intensity for protons of 500 MeV in Villigen (AG) (High-Energy Physics)" – its official title – 92.5 million francs.

This budget had been prepared by the middle of 1963, primarily by Willax, but also by Architect Schindler of Zurich for the buildings, in cooperation with the Federal Construction Office. The physicists submitted a restrictive budget. President Pallmann was clear on how challenging this project would be. He wanted to make sure that sufficient funds would be available, especially for the construction of the innovative equipment for experiments. He, therefore, made sure to include sufficient reserves in the budget. Additionally, the expected cost increases by 1965 were included.

Building Budget Plan 1965

Total credits were divided as follows (in million francs):

Construction of the proton accelerator	*30.0*
Housing for accelerator with halls for experiments	*20.0*
Operations building	*8.2*
Administration building with lecture hall	*2.5*
Laboratory building	*2.1*
Heating facility	*1.1*
Air raid precaution	*0.7*
Electrical equipment	*5.9*
General technical equipment	*10.5*
Site improvements and landscaping	*3.8*
Garages and bus stop	*0.3*
Apparatus, machine tools and equipment	*4.0*
Furniture	*0.5*
Unforeseen expenses	*2.9*
Total	**92.5**

The meson factory was built on a "Greenfield site" (see fig. 10). The site improvements included a bridge across the Aare River leading to the already existing Federal Institute for Reactor Research (EIR). This allowed joint usage of infrastructure belonging to EIR.

The cost for the beam facilities (proton beam, targets, beam dump) to be developed and adapted to each research project, were financed by appropriately generous annual equipment credits (in the first budget of SIN of 1974, they amounted to nearly 13 million francs). The costs for shielding the beam facilities were included in the housing budget.

Not included in the above budget, but mentioned in the budget plan, were the costs of acquiring the unimproved property of 1.5 million francs, which had already been concluded, as well as the costs – mostly salaries – of the meson factory development to date. These costs were financed by the operating budget of ETH.

For reasons of transparency, the expansion of existing buildings and the construction of additional ones were already mentioned in the budget plan, but application for credit was postponed because of the need to wait for requirements to develop.

For the later operation, the budget plan included about 200 qualified team members, which would cost between 7 and 10 million francs annually.

The funding proposal gives as reasoning for the new facility essentially the arguments described under paragraph 2.2. When one considers that this text was addressed to parliamentarians, who were hardly familiar with the subject matter, it was a masterpiece of general comprehensibility. The text also describes the expected diverse scientific usability: Advancement of nucleonic and meson physics at low energies, especially considering the kind of precision measurements for which no appropriate accelerator exists; nuclear physics at high energies – an area that would break new ground; new applications in radiation medicine and biology; manufacturing of special isotopes for chemistry, biology, technology; tackling problems of solid state physics and technology. It was also mentioned that the specialists at ETH and their international advisers had already prepared a diverse list of important experiments for the utilization of the facility.

The funding proposal also explained the reasons for the location near the EIR. Alternatively, the ETH satellite in Höngg could have been chosen. In favor of Villigen was the purely "industrial" character of the facility which would have been a poor fit on a campus. Additional important considerations were the safety aspects (fencing and radioactive shielding), space requirements, accessibility to cooling water and to a high-voltage power line of the electricity company Nordostschweizerische Kraftwerke AG (NOK), as well as the proximity to EIR with its existing infrastructure. Without being mentioned, it was logical that a Villigen location would be considered more "neutral" by the universities than would have been a location owned by ETH.

Fig. 10: Construction site on the western bank of the Aare River December 1968

All of this sounded very optimistic – and the later development would justify such optimism. But for a long time this project was not undisputed among physicists. After the funding proposal had already been prepared, another dramatic situation would develop. More than fifty renowned personalities from academia and politics signed a letter addressed to the Federal Council demanding abandonment of a meson factory for science-political reasons. It was assumed that the letter had been orchestrated by the already retired Paul Scherrer. Hans-Peter Tschudi of the Federal Council, however, was hardly impressed. He mentioned to Blaser that if so many personalities were opposed to a project, it was proof that it had an important role to play. Blaser should continue. From the standpoint of science policy the meson factory was already secured. As was mentioned in the funding proposal, realization of the project was supported in May 1965 by the Swiss Scientific Council, the Federal Council's most important independent advisory body.

The funding proposal was accepted in March 1966 by the Swiss Parliament. Unfortunately, the ETH Board President Hans Pallmann, who had pushed through the development of physics at ETH with both plans for the Höngg campus and the meson factory, would not live long enough to witness this parliamentary success. He passed away shortly before it happened. The Board's vice president, Claude Seippel, who was also supportive of the meson plant project, took over temporarily. In 1966, Minister Jakob Burckhardt of the Diplomatic Service was elected president of the ETH Board and would continue in that role until 1978.

The budgeted 92.5 million francs would soon be supplemented by a contribution from the Swiss National Science Foundation. It was obvious that the above-mentioned budget did not include an amount for the injector. A report dated summer of 1970 concerning the future tasks of SIN mentions 9 million francs which are dedicated for the construction of Injector I for the benefit of the universities. During parliamentary debates regarding the additional funding (see below), 4.7 million francs were mentioned.

In their 1972 ETH building budget proposal, the Federal Council applied to the Parliament for additional funding of 34.45 million francs, based on inflationary cost increases, for SIN which was in the process of being developed at that time. In their Winter Session 1972, the Council of States decided to eliminate the funding for SIN. This action had been led by Basel State Councilor Willi Wenk because an Injector II had been mentioned in the documentation. Funding for this injector was not part of the funding application, but SIN had announced at the same time that they intended to build a new injector. What obviously bothered Wenk was that, together with an application for additional funding for the completion of the SIN facility, it was already being considered in need of improvement. During a hearing conducted by a National Council Committee at SIN, Director Blaser declared that Injector II was not a subject for discussion. The intention was to finance it partially from SIN's operating budget (1 million francs annually) and partially through savings on additional facilities.

Federal Councilor Hans-Peter Tschudi, supported by the Federal Finance Department (which was represented by Vice Director Hans-Ulrich Ernst in SIN's building committee), was strongly in favor of granting the additional funding, as evidenced by the following citation from the Bulletin of Federal Councils:

"The Finance Department is able to judge this request for funding because management of the Finance Department is involved in the building committee for the Swiss Institute of Nuclear Research. I must emphasize – as Mr. Wenk has already mentioned: The Finance Department is of the opinion that an exemplary facility is being built here which can serve as a model for others in regard to planning and cost-efficient construction, meaning that the best report card can be given with respect to cost efficiency, good construction organization and excellent construction planning by the Building Commission under the leadership of former Brown Boveri director, Mr. Seippel, PhD."

In spite of such a strong testimonial, the Council of States denied the additional funding. However, during the spring 1973 session of a Federal Commission under the leadership of Mrs. Lilian Uchtenhagen, it was supported. Additional support came from parliamentarians from French-speaking cantons active in science policy who cited positive statements made by CERN in regard to the SIN facility. Especially Geneva's State Councilor Olivier Reverdin, who also presided over the Swiss

National Science Foundation, was in favor. This would indicate that the directors of SIN, who had been active in science-political circles, were being successful. The National Council unanimously passed the request for SIN funding and passed the matter back to the Council of States. State Councilor Wenk remained opposed, but the tables had been turned so that the Council of States passed the additional funding in March 1973.

Therefore, the early investments in the SIN facility (including site acquisition and contribution from the national fund) ran to about 137.5 million francs.

3 The Founding of SIN

3.1 *From Gloriastrasse to Oerlikon*

Up to the time the funding proposal for SIN was approved, the cyclotron planning team was located in two places. The physicists conducted their theoretical and computational work in the physics building of ETH on Gloriastrasse, while research laboratories were available to them on Universitätsstrasse. The team consisted of about 20 people in 1963. They worked on the 2-stage concept developed by Hans Willax in 1962 which consisted of an injector cyclotron and a ring cyclotron, and they were planning on acquiring the injector cyclotron from a commercial source. The development of the accelerator facility took several years, and adaptations to the concept were being made up to 1967. Top specialists were being consulted. Thanks to his former presence at the cyclotron facility in Berkeley, Willax was successful in gaining several experienced American engineers for the early phases of the project.

A Committed Team

The former cyclotron planning team consisted of a committed group of mostly young physicists. The supervisors, Jean-Pierre Blaser and Hans Willax (a dyed-in-the wool Bavarian with his accordion) did a lot to ensure that friendships were cultivated even outside of work. Among their activities were regular bowling matches at the Restaurant Rigihof near the ETH Fridays after work; the traditional May walk with a barbecue in the forest; skiing weekends in Les Diablerets and on the Lenzerheide, dormitory-style, of course; raclette grill parties at Blaser's home; and finally, soccer matches between the different departments.

Additionally, Director Jean-Pierre Blaser commissioned the American firm William B. Brobeck to monitor the SIN project. A report entitled "A Survey of the Ring Cyclotron Project" was published in July 1968. It contained exact information regarding costs, personnel requirements, time line and space requirements. A total cost of 97 million francs had been calculated for the project, as well as a staff requirement of 195 at full operation, 155 for research.

One has to remember that the participating physicists did have the self-confidence in their ability to build a machine in accordance with the appropriate specifications. However, many problems of physics and technology still had to be solved on the way to success, and that still left the possibility that the concepts would not work as envisioned. For instance, at the outset it was uncertain if the accelerator cavities would be able to maintain a voltage of 300 kilovolts. In spite of the fact that most of the staff had little experience with particle accelerators, Jean-Pierre Blaser had full confidence in his young physicists. However, to assure himself to a certain degree, he would probe his team now and then by sending through experts disguised as interested visitors.

To the ETH Board and other Federal Authorities, it went without saying that they could not just leave the realization of such a large and technologically ambitious project to a small group of physicists with no industry experience. That is why they entrusted the project management to Maschinenfabrik Oerlikon (MFO) which also supplied the large magnets. The division of labor was formulated as follows in the funding proposal of 1965:

"The proper solutions to the many different development tasks require mathematics, physics and engineering knowledge in equal measure so that they fit well into the responsibilities of a technical university. However, the realization of the machine requires knowledge of construction technology, organizational and commercial tasks to such an extent that they can only be handled by an efficient industrial organization. Therefore, the construction technology, the division of labor and the commercial transactions for the project will be handled by Maschinenfabrik Oerlikon (MFO). According to the contract, the planning team of the ETH is responsible for the facility concept, for drafting the specifications and for the development of the basic technical solutions. The MFO will act much like a general contractor who handles the technical construction tasks and ensures completion in accordance with specifications."

According to the funding proposal, the time line for the completion of the concept was pretty complicated. Exact dates had to be set for the construction of the experimental hall, the utilities and the development of the machines, and all had to be synchronized. Often the buildings had to be constructed to exact measurements and to be built around the machines during installation.

The physicists and engineers agreed to the following time line: The machine development, including extensive model testing of magnets and high-frequency cavities, was to be completed 1965/66. During construction of the machine components (1966 and 1967) the large experimental hall had to be completed to the extent that the delivered machine components could be stored, checked and pre-assembled. The actual assembly of the ring accelerator had to start in 1968, and the injector had to be commissioned. Testing of the complete system could then be expected to start in 1969. Right in the middle of the development phase (1966-1967), active preparations for the research needed to be started with the collaboration of the appropriate university institutes. As it turned out, this time line proved to be too optimistic: The SIN facility did not produce its first π-mesons until 1974.

The personnel head counts also had to be adjusted to the respective development and building phases. The development group of ETH needed about 40 employees, half of them physicists or engineers. Added to that was the coordinating group of the MFO as well as construction and installation personnel made available by the company – 15 to 20 people on average.

Fig. 11: First magnet model 1:5 scale at the MFO, fall of 1964. First on the right Jean-Pierre Blaser, next to him MFO Director E. Schnorf.

The collaboration with MFO from 1966 on had a decided advantage. The development tasks and first assemblies of the large machine components could absolutely not be handled in ETH laboratories, and it would take more than half a decade before the new buildings in Villigen could be occupied. Therefore, MFO's invitation to occupy barracks on their "Stierenried" property in Oerlikon and use the available hall for assembly and tests of large components was most welcome. That made it possible for Willax and his growing development group to move there and work under optimal circumstances. The fact that the project phase at MFO worked out so beautifully was due in large measure to the kind assistance of one of the company's directors, E. Schnorf.

Especially the unique sector magnets, which were built by MFO, could be assembled and measured in the halls. The complex shape of the magnets and the high demands on accuracy involved in an isochronal cyclotron – at a time when computers were rarely in use – made this a challenging task. Here the collaboration of MFO with their testing facility proved to be most fruitful. By the way, for the work on the magnet gaps of the sector magnets, Switzerland's largest lathe was needed, which was located in the Birrfeld plant of Brown, Boveri & Cie (BBC).

Fig. 12: Work on the magnet poles of the 1:1 magnets at BBC Birrfeld: Attached pole pair on the 14-meter large carousel lathe (rough-machining of pole areas with roughing tool), fall of 1968

Another challenge was the high-frequency system for Willax's sector cyclotron. The several meters high cavities with their gaps for the particle beams required support structures in order to withstand the enormous vacuum forces and had to be made of a highly conductive material. The accelerator builders chose aluminum, and it so happened that a company with initiative and interest in solving difficult problems was located nearby – the Aluminium-Schweisswerk Schlieren (ASS) – a company with which the group developed a long-term and very successful collaboration. For lack of good outside offers for the high-frequency emitters with high capacity, the decision was in favor of in-house development under the direction of Paul Lanz.

Fig. 13: Cavity before insertion into the vacuum chamber, MFO Hall, fall of 1965

The partnership with MFO also had its disadvantages. Their planned collaboration in handling the technical-administrative tasks of the project quickly turned out to be illusory. It became obvious that important decisions could only be made by specialists who had an overall understanding of the entire facility. An unusual number of partial aspects in the development of this unique accelerator were closely interrelated: Very large magnets with the appropriate mechanical forces; electrical feeds as well as extensive cooling systems in the magnitude of several megawatts; the creation of the necessary high vacuum inside the accelerator and its beam channels; the path dynamics of the atomic particles which would fly through the accelerator and beam channels; electromagnetic high-frequency fields with very high power; strong shielding measures necessary to protect from the radioactivity created in the components, and last but not least, a control system for the entire installation that was based on measured and target parameters.

The specialized mechanical and electrical engineers of the MFO lacked the necessary overall understanding so that the physicists involved ended up making the decisions. The MFO specialists also had more of a braking effect when it came to unconventional ideas which the mostly young and impartial physicists had all the time. For instance, the MFO engineers insisted that the superconductive magnet channel for muons developed by Georg Vecsey would never work – a position for which he later provided evidence to the contrary. The engineers had assumed that the channel would be cooled by a bath of liquid helium which would evaporate during cooling and let collapse the superconductivity. Vecsey solved the problem by pumping the highly critical helium through the magnets under high pressure.

For all these reasons, a new contract was concluded in 1970 between the Confederation and MFO – which had been acquired in 1967 by BBC – that released them from their responsibility of acting as a "general contractor", as agreed to in 1964, and instead assigned to BBC/MFO precisely defined technical and planning responsibilities in regard to the ring accelerator.

3.2 *The Beginning of Construction in Villigen*

After the Office of Federal Buildings had already contracted with architect Schindler for the design of the buildings described in the funding proposal, this company also received the order for the buildings in Villigen. This turned out to be extremely fortunate because the facility was consistent with an industrial site and required the types of technical competencies that Schindler offered. Schindler had made a name for himself with a project in Zurich that he completed overnight: Building the access bridge across Rämistrasse to the parking structure "Hohe Promenade".

The collaboration with Architect Schindler started out amicably and was decisive in the success of the SIN facility. The first building project in 1969 was the bridge across the Aare River that connected EIR and SIN. This bridge was essential for regional access to SIN, whose users were expected to come from universities all over Switzerland, as well as for the desired collaboration with EIR. In addition, it served important supply lines, such as for potable and cooling water. Because of the expected heavy-weight traffic, it was designed for an average load of 75 tons. September brought a spectacular load test in the form of a military transport of 105 tons.

Fig. 14: SIN construction site, October 1969: Bridge across the Aare and building excavations for utility building, experimental hall and operations building.

Construction of the buildings in Villigen was a demanding task, and not only because of the previously mentioned time line. The ring accelerator was just in the development phase. Therefore, its final dimensions, upon which to calculate the foundation for the heavy-weight components, were not yet available, and neither were its technical specifications upon which to calculate electricity and cooling requirements. Additional difficulties presented themselves in the design of the large experimental hall, making it sufficiently flexible so that it could accommodate experimental areas that were still in development with their many magnet-focused particle beams, radioactivity shields weighing several tons, necessary lifting equipment – central to which was a lifting crane with a 60-ton capacity – as well as the electrical and cooling water channels, all of which had to be adaptable to future requirements. In spite of everything, Schindler managed to complete the buildings on time and on budget, which gained him the respect and gratitude of the physicists.

Fig. 15: Casting concrete for the foundations of the ring accelerator, fall 1969

In December 1970 the SIN area in Villigen looked as follows. The experimental hall – unmistakably the central piece of the facility – was still under construction. Already mostly finished were the transformer station as well as the utility building with its ancillary structures for electricity and cooling. The first cooling used ground water. For the multiple water pipes – cooling water, potable water, and waste water – Schindler had designed a well-dimensioned system of underground channels. A waste water treatment plant had also been completed. The operations building was also well on its way. Its equipment, as well as the equipment for the laboratory building, had been adapted to requirements which by now were more easily assessable.

3.3 *SIN as an Annex Institute to the ETH*

On January 1, 1968 SIN was dedicated as an annex institute to ETH. Jean-Pierre Blaser was elected as its first (and only) director. The dedication took place legally through the enactment of a Federal ordinance that specified SIN's mandate and primary agents. The ordinance mentioned the function of the user laboratory for the benefit of the universities. It also contained the responsibilities of its director as well as the primary elements of the organization.

In an introduction to the first annual report of SIN – which in addition to "SIN" was also referred to as "Cyclotron Planning" – Director Blaser thanked the late ETH Board President Pallmann without whom the project could never have been realized: "He had recognized the important function of the undertaking for ETH and considered the creation of a national research center an important science policy factor for our country. He was especially happy about the versatility of its research facilities which furthered the dialog between the disciplines – physics, technology, chemistry, medicine and biology." Blaser continued: "Built on trust by ETH, SIN will be open to all Swiss universities. International participation is also expected because the synchrocyclotron of CERN will make way for SIN with its significantly higher energies."

The decision to found SIN was, without a doubt, to the benefit of Switzerland. The institute had a clear focus. It served the needs of Swiss universities in various areas, especially since its facilities offered broad applications. Therefore, several scientific disciplines were assembled. Worldwide it had only two competitors: The meson factories LAMPF in the USA and TRIUMF in Canada. Thus, it was unique in Europe and gained international interest.

But it was not just the underlying conditions that were favorable. Also favorable was the way in which the director organized the leadership together with his management team. SIN management led the organization in an entrepreneurial fashion, yet like a university institute, and valued strong connections with Swiss universities. SIN had several infrastructure possibilities of furthering the universities. And those, in return, supported the institute on a scientific level.

The construction budget approved for SIN by the Confederation was temporarily administered by the Office of Federal Buildings. A so-called Building Committee oversaw the project managed by Jean-Pierre Blaser. The ETH Board had assembled the following members for this committee: Claude Seippel (ETH Board and BBC) as president; Konrad von Wurstemberger (ETH Board) as vice president; Hans-Ulrich Ernst (Federal Finance Administration); Professor André Gardel (EPFL); Dr. Alexander Goldstein (BBC); Fritz Grütter (BBC); Hans U. Hanhart (regional director of the Office of Federal Buildings); Professor Urs Hochstrasser (Federal Office of Science); Professor Verena Meyer (University of Zurich) as members. At the end of 1974, with the beginning of the research operation, this committee was dissolved.

When SIN was founded in 1968, it consisted of just a project team and a staff of about 100 people. The future location of SIN had been determined, but the buildings still had to be constructed, while the employees were working on the site of MFO in Oerlikon. In the Swiss State Calendar, only the "seniors" (permanently employed because of their age) Willax, Baumann and Lanz are listed.

SIN issued its first annual report in 1969. It contained an organization chart referred to as "provisional" (see Addendum 2): SIN consisted of the two departments "Accelerator Facilities" under the direction of Hans Willax and "Research" under the direction of Hans-Jürg Gerber. There was a Staff Division under Wilfred Hirt, who was responsible for the coordination between research and machine and later became Deputy Director.

As the organization chart shows, the organizational structure was not yet final. It was concentrated on the specialists who had participated in the meson factory project back in the 1960s. Many of the planned supervisory positions were still unfilled. Naturally, the organizational development of Hans Willax's department was further along since the first priority was the construction of the machinery, while the research that would use the machinery would follow later. In addition, the organization chart listed especially those persons who, because of their functions, were outside partners, such as Federal Administration, construction companies, vendors.

The internal partner handling the SIN project for Architect Schindler was Theo Schaub. The small but efficient administrative group was overseen by Max Werder. It received notable support from the Federal Administration of Bern. Especially attorney Dr. Ernst Schneeberger of the Federal Finance Administration, who was responsible for the Federal Administration's procurement division, advised SIN's Administration on the country's extensive and complicated procurement procedures (and later insisted in 1988 to introduce the new head of administration of PSI to the country's finance and procurement regulations, although he was retired by that time).

Hans Willax's department, which built the ring cyclotron, consisted of the three sections Accelerator Development, Mechanical Engineering and Electrical Engineering. The Accelerator Development Group consisted of Werner Joho (beam dynamics), Branko Berkes (magnets), Paul Lanz (high-frequency system), Urs Schryber (accelerator physics) and Ludwig Besse (controls). There was also key personnel without supervisory function, such as Miguel Olivo (extraction elements and beam diagnostics). The Mechanical Enginerring section was also headed by Theo Schaub. This section consisted of the groups of Jürgen Langhans (construction), Hans Frei (vacuum) and Hans Oschwald (shop floor). In the Electrical Engineering section under Hans Baumann, the group leaders were Gottfried Irminger (supply) and Alfred Boss (electronic laboratory).

In Hans-Jürg Gerber's department, which was in charge of equipping the experimental hall and constructing its areas, only few of the planned supervisory positions were filled: Wilfried Schoeps (instrument electronics), Charles Perret (radioactivity shielding, activation, radiation control) and Janos Zichy (beam optics).

Fig. 16: SIN site, construction of the experimental hall, transformer station, and the operations and the utility buildings, August 1970

In 1970 a new organization chart was created with the support of external advisors, partially because of increasing personnel and partially because of the upcoming integration into the Federal Administration (see Addendum 2). This chart divided the departments into Accelerator, Experimental Facilities, Technical Services and Administration. Hans Willax was the new deputy director. Now there was a theory group under the direction of Florian Scheck. Hans-Jürg Gerber's department was now more precisely called Experimental Facilities. New names in this flowchart were: Christoph Tschalär (apparatuses), Marc Pepin (research group and computer facilities), Reinhard Frosch (secondary beams), Gilbert Guignard (beam controls), Georg Vecsey (cryogenics) and Salvatore Mango (polarized targets). Again, there was also personnel without supervisory function, such as Claude Petitjean, Erich Steiner, Manfred Daum and Hans-Jürg Leisi. New in Construction was Emil Mollet (planning and building supervision), and in Administration there were Cyrilla Kijewski, responsible for Human Resources and Josef Nussbaumer, responsible for Accounting and Procurement.

Fig. 17: Casting concrete for the foundation of the ring accelerator, two thirds complete in October 1970

The organization of SIN functioned under the following three characteristics: First of all, Director Blaser believed in a cooperative leadership style. The executive committee was the so-called directorate. It consisted, in addition to the directors, of department managers, executive assistant as well as the head of theory. The cooperative leadership style filtered downward. The professional opinions of staff were valued, making staff members act with confidence. Therefore, identification with the institute was accordingly high.

Secondly, supervisory positions were generally not desired. Especially the young physicists were more interested in their fields of expertise than in a supervisory function. Blaser and Hirt were of the opinion that the best supervisors were those who had been urged into their function. The success of SIN showed that they were right, in spite of the fact that this philosophy sometimes had the result of leadership being neglected. Meanwhile, Hirt especially was well-informed about personnel goings-on and did not hesitate to replace supervisors who had proven unsuitable.

Thirdly, the "static" structure of the organization chart was overlaid with a dynamic one in that often ad-hoc groups were formed for partial projects which consisted of physicists, engineers and specialists in various departments. Qualified employees were named project leaders of non-hierarchical study groups.

For the most part, the above-named personnel remained with the institute for years as valued knowledge carriers because the development, operational optimization and further upgrading of the facility were all long-term projects. In the coming years, the SIN team was expanded further, and the institute was able to attract very capable and proven specialists in all areas.

Fig. 18: 60-ton lift crane for the experimental hall prior to installation, November 1970

The work of Wilfred Hirt, who was named deputy director in 1977, deserves special mention. Blaser and Hirt made a strong team for many years. Their personalities complemented one another beautifully and they trusted each other implicitly. The most important decisions were made jointly.

Wilfred Hirt

After completion of his dissertation at CERN, Hirt joined the management team of SIN. He devoted the following years to the development of the organization. Time and again, management challenges had to be solved for which he displayed a special talent. He had an excellent feel for the strengths and weaknesses of employees. In many conversations with them he was able to reach consensus and could take appropriate measures. He took part in the development of SIN as an Annex Institute and introduced many policies which established the institute as a national research center and user laboratory. In addition, he was responsible for the management of human resources and finance. Here he often chose unconventional approaches. However, his style was received favorably at the Federal Administration and allowed SIN to dispense with a highly developed bureaucratic hierarchy.

Hirt played a decisive part in overcoming political and academic attacks which arose during the approval process for the Injector II and the spallation neutron source. Although he was responsible for management tasks, he regularly took part in physics colloquia. He conceptually shaped the international collaborations of SIN as well as the initiatives of science policy, for instance, the continuation of the accelerator mass spectrometry at ETH in Höngg, or as manager of the acquisition of the RCA laboratory in Zurich. His initiative was decisive in the preparations for the merger of EIR and SIN into PSI. Many of his personal contacts led to the germination of this idea, and when the ETH Board asked Hayek Engineering to hammer out the details, Hirt was the actual project architect.

4 Construction of the Accelerator and Research Facilities

4.1 *Concept and Machinery Arrangement*

As Jean-Pierre Blaser wrote in his introduction to the first annual report of the institute which was published in 1969: "Our goal is the establishment of a research facility in Villigen which is equipped with a 590 MeV accelerator for high proton currents [100 microamperes] and ancillary equipment for experimental research in the fields of elementary particle and nuclear physics."

In the funding proposal, the proton energy had still been described as 500 MeV. After measurements at CERN regarding the rate of pion production, the target energy for the ring accelerator was increased in April 1969 – practically at the last moment – from 500 MeV to 590 MeV. This required not only new optimization of the sector magnets, but also an increase in injector energy from 68 to 72 MeV.

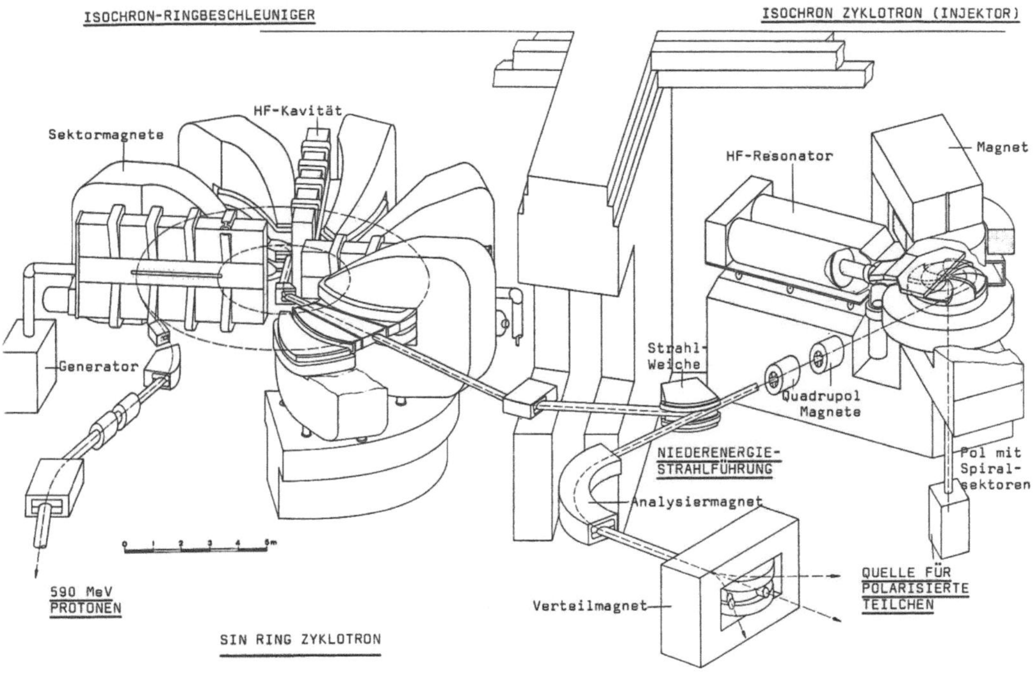

Fig. 19: Schematic of the 2-stage accelerator facility at SIN, with the Philips injector (right) and the ring accelerator (left)

In June of 1969, the developers started testing prototypes of machine components in the halls in Oerlikon. The injector for the first acceleration phase turned out to be a special case. Under consideration as suppliers from industry were Philips of the

Netherlands, AEG of Germany, CSF of France and the Cyclotron Corporation of California.

Philips and AEG responded to the request for quote. AEG opened the poker game by guaranteeing a beam current of 100 microamperes. On the following day, Philips was forced to increase their offer from 50 microamperes to 100 microamperes if they wanted to remain in the running for the order. In addition, the Californian Cyclotron Corporation offered a cyclotron with negative hydrogen ions – analogous to the one at TRIUMF – which they introduced at an industry exhibit in Basel. At that exhibit, a representative of the company besieged the visiting Werner Joho to please convince Blaser that their machine was the best one of all! As described under paragraph 2.3, the developers of the accelerator would have preferred the machine from AEG because it promised a better capacity of the whole installation. Under consideration of the support of SIN by the universities, Blaser and Willax decided on the injector from Philips. It was ordered in the fall of 1968 after the conclusion of tough negotiations. Shortly thereafter, Philips began building the injector cyclotron.

Fig. 20: Foundation for the ring accelerator in the ring bunker in December 1970

The machine components for the ring had been ordered or were still in development. Team members often came up with brilliant solutions, such as the idea of vacuum engineer Hans Frei of a torus-shaped vacuum seal between sector magnet and high-frequency cavity which enormously eased the move-out of a cavity during an installation revision.

Construction at the Villigen site progressed according to plan. The time line was now considered secure. Experimental operations were scheduled to begin in the middle of 1974.

Res Jost, Professor of Theoretical Physics at ETH, suggested the formation of a Theory Group to accompany Research. Jost had already influenced the ETH board president regarding a shift from nuclear to particle physics, which he considered very promising. He also wanted to include these new scientific questions in his curriculum. In response to Jost's suggestion, SIN management entrusted the young theoretician Norbert Straumann with the establishment of the group. The many lectures and courses given by Straumann are well-remembered by the attendees. When Straumann decided to change his field of activity, Florian Scheck took over leadership of the Theory Group in 1970. This group was very active and was the reason for many visits by important scientists. The group also proved highly valuable in the judging of suggested experiments.

4.2 *Arrangement of the Entire Facility*

For the construction of a meson factory it was not enough to develop the accelerator which would supply the intensive proton currents. Concepts had to be developed and realized which would cause the protons to produce intensive pion beams with which experiments could be conducted.

In response, the builders of SIN developed the following concept: The protons extracted from the ring accelerator are being guided into a proton channel and would bombard so-called production targets, thereby creating pions through a nuclear reaction. The proton beam would dissipate in the air after a short distance because of its diffusion with air molecules. Therefore, it needed to be transported through a vacuum tube. Because of the so-called beam divergence, the beam expands during transport, even in a vacuum. The guided beam has to be constantly squeezed together again with the help of quadrupole magnets. Diversions of the proton beam are also created with magnets.

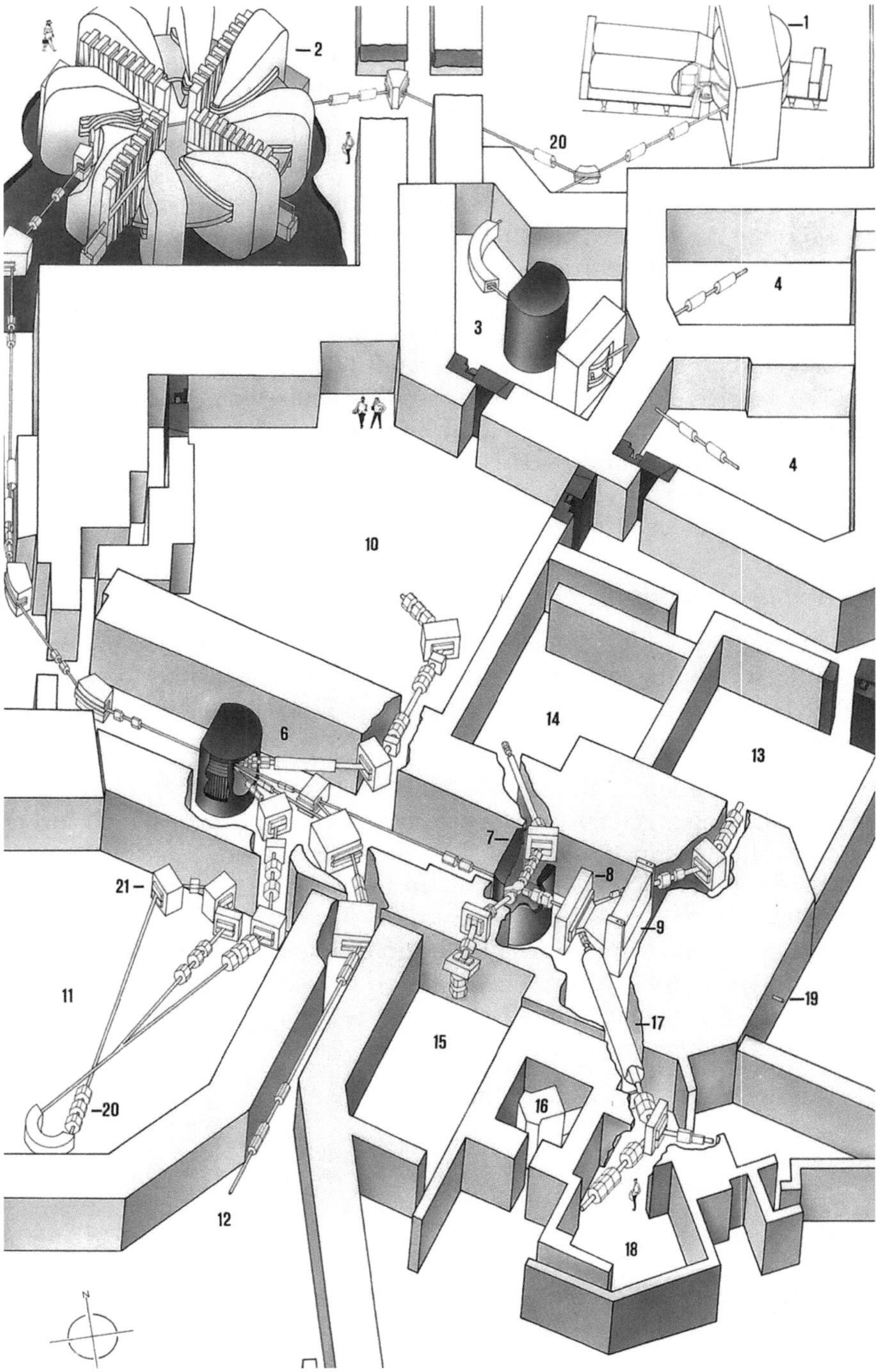
1
2
20
3
4
4
10
14
13
6
7
8
9
21
11
19
17
15
16
20
12
18
N

Fig. 21: Schematic of the Experimental Hall

1 Injector cyclotron

2 Ring cyclotron

3 Analysis bunker

4 Low-energy experimental areas

5 Extracted proton beam

6 Target station M (production of pions and polarized protons)

7 Target station E (production of pions and neutrons

8 Diversion magnet

9 Beam collector

10 Experimental area πM1 (high-resolution pion beam)

11 Experimental area πM2 (pions of high energy and good beam quality)

12 Experimental area pM1 (polarized protons)

13 Experimental area πE1 (intensive high-energy pion beam)

14 Experimental area πE2/μE4 (low-energy pions and short muon channel)

15 Area for research in biology and medicine

16 Crystal spectrometer

17 Muon channel (super-conductive solenoid)

18 Areas μE1 to μE3 for muon experiments

19 Shielding wall

20 Quadropole magnets (electromagnetic lens)

21 20-ton bunker door

The beam guidance quality was of primary importance to the success of SIN as a meson factory. Protons which leave the beam create radioactivity in the channel walls. During operations, this radioactivity is shielded by walls of steel and concrete blocks, however, it should not exceed certain limits after a falling stage because it would render service and repair work practically impossible. It was only because the developers of the SIN facility succeeded in significantly reducing proton losses that, over time, they were able to increase the proton beam to today's values – which put them at the top of meson factories worldwide.

The main proton beam is first fired on a thin target and produces pions. This target is known under the abbreviation M (from the French "mince" – in the German as well as the English languages, the words "dick, dünn" or "thick, thin" cannot be distinguished by differing first letters). Most of the protons leave the target M and are guided into target E ("epais") where, again, they produce pions, muons and neutrons. The remaining beam was caught in a beam dump – it feeds the spallation neutron source today.

The development of both target stations, for which Christoph Tschalär was responsible, posed a particularly difficult challenge. The proton beam creates significant heat inside the targets. Without a cooling mechanism, the materials would evaporate. The problem was solved by letting the beam tangentially touch a saucer-shaped target which constantly rotates around its axis so that the heat is dissipated over the entire target. The target mechanism contained four such wheels which were made of graphite, molybdenum and beryllium.

Fig. 22: Target head with four target wheels

By being bombarded with protons, the target wheels became highly radioactive. Therefore, larger revisions had to be conducted in a hot cell. The construction was designed in such a way that the targets could be replaced by remotely controlled and shielded so-called target replacement casks, without exposing personnel to radioactivity. Another problem arose with the turning mechanism of the target wheels. The high-intensity particle irradiation precluded the use of conventional lubricants. Tschalär was finally able to solve the problem through appropriate surface treatment and by using molybdenum sulfide.

Fig. 21 shows the arrangement of the experimental hall with the listed facility components. Beginning at target M, three different pion beams as well as a beam of polarized, diffused protons are guided into four experimental areas. Beginning at target E, it was three pion beams, one divided beam of muons as well as two neutron beams for a total of seven experimental stations. One pion beam was reserved for biology and medicine. It was the only one of these secondary beams that was aimed downward – in consideration of possible radiation treatment of patients. Reinhard Frosch, together with Erich Steiner and Manfred Daum, was responsible for the development of these secondary beams. The injector cyclotron also delivered various beams for experimentation in two areas. Their dimensioning was the responsibility of various university groups, but the accelerator department built and operated them.

Fig. 23: The SIN site in January 1971. Seen on the site is the large experimental hall with saw-tooth roof, flanked on the Aare side by the operations building, on the mountain side by the utility building; in front of the facility grounds the transformer station and at head of the bridge leading to EIR, the waste water treatment facility; the laboratory building under construction alongside the access road and adjacent to it the site for the assembly hall.

Several megawatts of electric power were needed for the operation of the SIN facility. Several hundred kilowatts went into the proton beam alone – today it's even more than 1 megawatt. The necessary power was delivered through the transformer station directly by the high-voltage network of the Aargau electricity supplier. The power distribution facility for getting the power to the users – high-frequency system for the acceleration, accelerator magnets and beam guidance magnets and cooling systems – was housed in a specially constructed utility building. The building also contains the extensive cooling systems which dissipate the heat through several cooling circuits finally into the Aare – and during the winter into the return flow of the regional REFUNA remote heating system. Since the cooling water for the innermost components is radioactive, tiered cooling circuits guarantee that no radioactivity is released into the environment.

Fig. 24: Still empty cable racks in the ring bunker, summer 1971

4.3 *Construction of the Facilities in Villigen*

On July 1, 1971 the experimental hall, the operations building and the utility building were occupied by their respective departments. Just as planned in the time line, the assembly of both accelerators began immediately. Since the planners had counted on many unforeseen circumstances, the most basic needed equipment had been stationed in Villigen already three months earlier. This advance planning offered much flexibility to task planning as well as the possibility of making quick decisions. In spite of several occurrences, which significantly influenced certain sequences of work, the scheduled program of machinery assembly for the rest of 1971 could be completed on time. Such occurrences were: the damage to a finished magnet pole during assembly in the shop; insufficient air conditioning for the computer room of the operations building with downtimes that delayed computer operations; the temporary breakdown of the 60-ton indoor crane caused by technical failures which could not be solved due to the tight time line.

A partial move of the Accelerator and Engineering Departments followed at the beginning of July 1971, but the majority of these departments still remained in Oerlikon. Regarding components, the capability improvement of the high-frequency generators, the development of the magnetic trim coils (with which the magnetic field of the ring accelerator could be evened out) and the development of the accelerator control system, were the top priorities. The physicists were also working on calculations for radioactivity shielding.

Fig. 25: Philips Injector Cyclotron

While working on this, the engineers and physicists also encountered partially unexpected difficulties. Continuing experiments with the high-frequency system caused overload on the tubes which lead to the destruction of their vacuum jackets; waiting for the development of improved tubes became necessary. Also the trim coil prototypes were unexpectedly heating up too much, which necessitated improved design and load testing on new prototypes.

Fig. 26: First sector magnet of the ring cyclotron, weighing 240 tons, in the hall of MFO, 1972

With the occupancy of the laboratory building in 1972, the move of SIN from Oerlikon, as well as from ETH to Villigen, was essentially complete. Only the High-Frequency Group remained in Oerlikon until it had completed testing the high-frequency generators in the spring of 1973. Nevertheless, 1972 would become a hectic year. Again, there were many unforeseen occurrences and meeting the scheduled time line was a top priority. Radioactivity shields were being built in the experimental hall, made of concrete and iron,whereby the latter had been made available from Swiss war material reserves.

Fig. 27: Cavity with vacuum chamber at the Aluminium-Schweisswerk Schlieren

In addition, the experimental areas described in the previous section were being prepared. Under the direction of Reinhard Frosch, beam guidance systems for the secondary beams – pions, muons, polarized protons and neutrons – had to be developed. Claude Petitjean was responsible for the muon rays. The super-conductive muon channel deserved special attention. The former (first) director of TRIUMF, John Warren, had warned Werner Joho during his stay in Vancouver in the early 1970s that SIN would take too large a risk by building the muon channel with a super-conductive solenoid, according to Georg Vecsey's plan. Uncertainties were the reaction of the solenoid under radiation as well as the cooling with highly critical helium. That was the reason why TRIUMF built its muon channel with the proven quadrupole magnets. Joho passed on these doubts to SIN, and also many specialists at CERN warned of this supposedly much too risky and allegedly unnecessary development. However, Director Blaser had full confidence in Vecsey and said: "If this new project works as expected, then we will be way ahead of the competitors!" And that's exactly what happened, and Vecsey's solenoid design would eventually be copied by all other meson factories.

Fig. 28: Installation of the Ring Accelerator, 1972

The development of secondary beams and experimental areas took place in collaboration with the users. A list of experiments being prepared, as well as their participating researchers and institutes, already existed. Because of the expected experimental operations, occupation of the assembly hall occurred on time in the middle of 1973.

Since public interest in the new institute was high, SIN management organized visitor services. This had to be a balancing act. On the one hand, feedback from the interested public was important to SIN, on the other hand, the challenging installation tasks could not be interrupted by visitor tours.

As planned, SIN was able to largely finish its accelerator facility in 1973, could test components and operating systems and get them functional for operations. The same went for the targets, the secondary beams and the experimental areas, although the installation of an enormous amount of shielding material took great effort.

Fig. 29: View of the almost completed ring accelerator

Fig. 30: (next page): The completed ring accelerator and parts of SIN personnel, September 1973.

4.4 *From the First Pion to the 10th Anniversary*

January 1974 brought proof that the injector cyclotron as well as the ring cyclotron had become reality according to their specifications. When tested, the Philips cyclotron delivered a beam of 100 microamperes, something the SIN physicists had not expected after the initial difficulties (routinely this value could only be reached in 1979, after SIN personnel had completed upgrades). During the following months, both machines were still getting optimized. In February, the 40-meter long, 590 MeV beam path to target M was ready to be put in operation.

Finally, February 24, 1974 was the big day: All the important physicists, Director Blaser among them, were assembled in the accelerator control room. At 12:35 a.m., right after passing through target M (the beam extracted from the ring hit target M without a lot of searching after a small magnet correction) a proton beam of 21 nanoamperes could be verified with the help of a scintillating fluid. Seconds after the proton beam was registered, a pion could be registered in the Pion Area No. 3, the one that was furthest along.

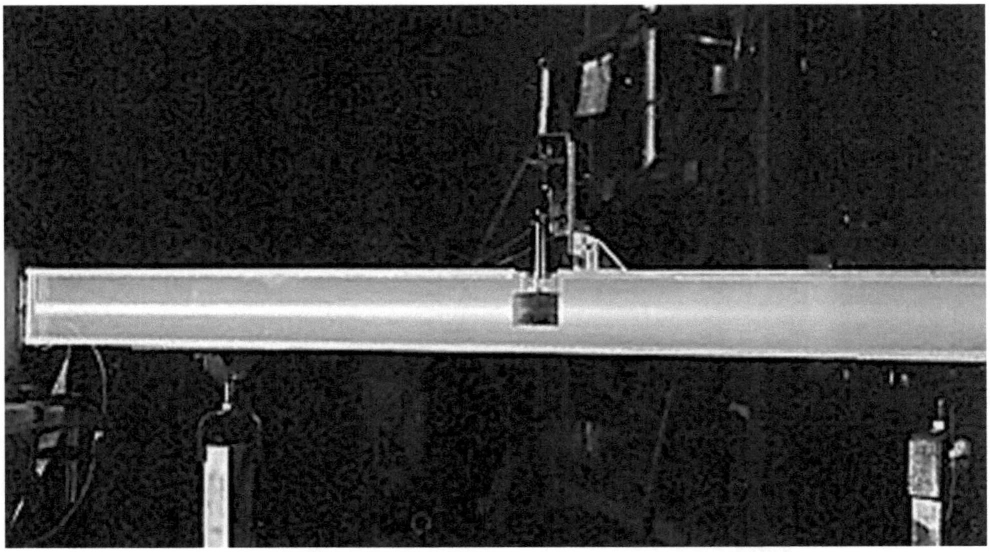

Fig. 31: The thick-as-a-finger proton beam, coming from the left, is made visible by scintillating fluid. It penetrates a copper block of 7 cm thickness

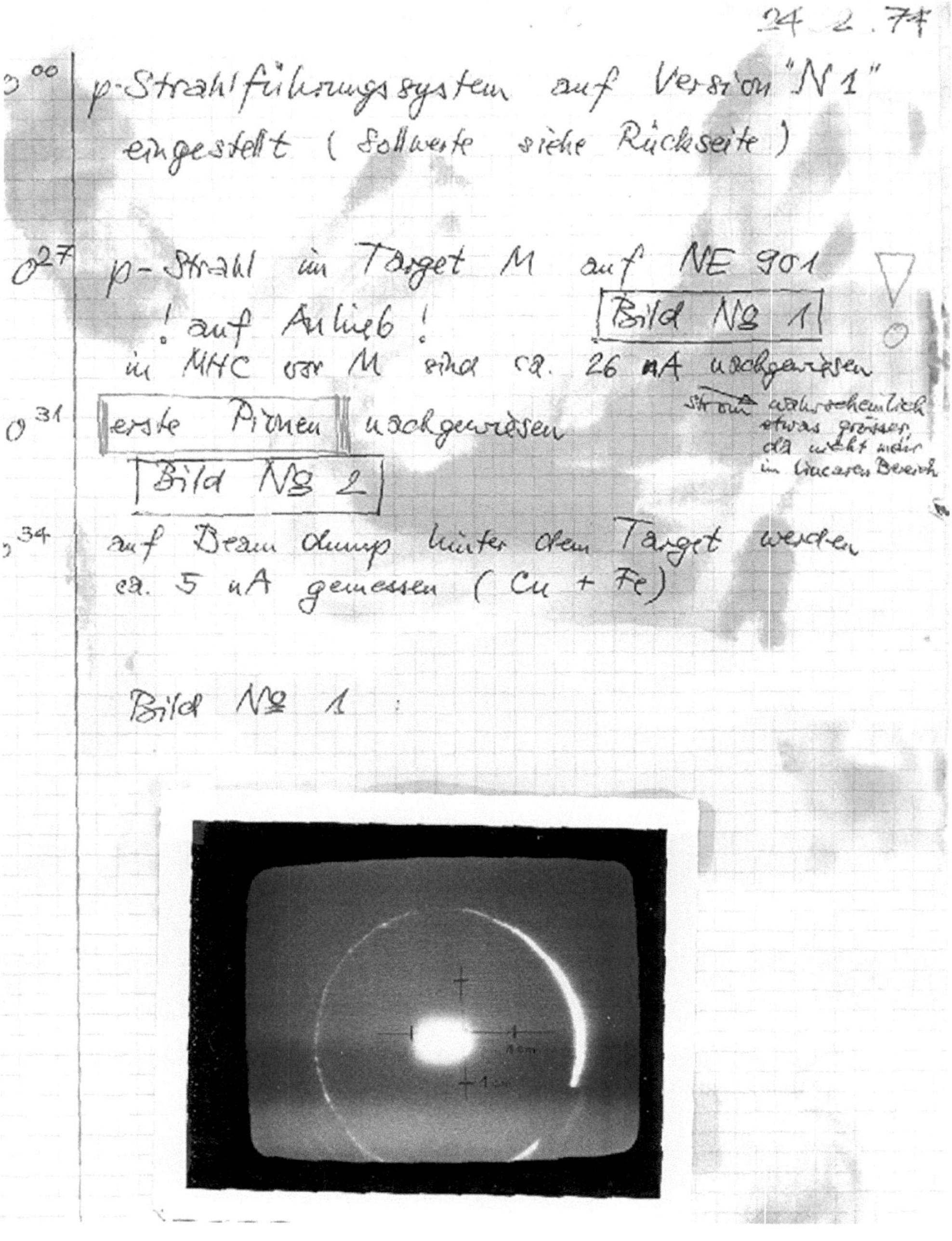

24. 2. 74

0 00 p-Strahlführungssystem auf Version "N 1" eingestellt (Sollwerte siehe Rückseite)

0 27 p-Strahl im Target M auf NE 901 ! auf Anhieb ! [Bild No 1]
im MHC vor M sind ca. 26 nA nachgewiesen
(Strom wahrscheinlich etwas grösser, da nicht mehr im linearen Bereich)

0 31 [erste Pionen] nachgewiesen
[Bild No 2]

0 34 auf Beam dump hinter dem Target werden ca. 5 nA gemessen (Cu + Fe)

Bild No 1 :

Fig. 32: Log entries the night of February 24, 1974

Fig. 33: February 24, 1974: First pions manufactured. In the control room from left Richard Reimann (all the way in the back), Thomas Stammbach (with pipe), Manfred Daum (partly concealed), Werner Joho, Franceso Resmini (all the way in front), Hans Willax, Paul Rudolf, Urs Schryber, Jean-Pierre Blaser. Not in the photo is M. Olivo. He later complained that all participants could be seen in the photo, except him – the photographer!

Director Jean-Pierre Blaser was able to determine that the "unexpectedly quick commissioning of accelerators and particle beams and the surprising operational reliability and stability" already permitted experiments. In the 1974 annual report he did one better: "The year 1974 is a pinnacle for the institute since the successful commissioning of the accelerators as well as the first scientific results fell into the same reporting year. It was a year of especially high tension for all participants who, after many years of single-minded development work, were facing 'the moment of truth' not without anxiety. Therefore, feelings of happiness and relief were great, but also feelings of justified pride when, very quickly and mostly with less problems than expected, all elements of the complex systems functioned well and combined to create a usable research installation."

The official dedication of SIN took place on October 4, 1974. Speeches were made by Federal Councilor Hans Hürlimann, CERN Director Willibald Jentschke, ETH Board President Jakob Burckhardt, the President of the SIN Building Committee Claude Seippel and SIN Director Blaser.

Fig. 34: Last portion of the proton channel during its development phase, 1974. Target Station M is in front, in which pions are manufactured by a collision of protons with carbon or beryllium. Target Station E (in the upper third of the photo) is still in development. Concrete blocks are erected all around the proton channel as massive shields against undesirable radiation.

Regular experimental operations started in 1975 for the first time; with regular breaks for the optimization and enhancement of the acceleration facilities (the so-called "beam development"). The optimization consisted primarily of the reduction of radioactivity to facilitate service work, and was also necessary for the realization of higher beam currents. At times this led to instability of the proton beam which limited the experimenters who wanted a stable beam, if possible.

In spite of disruptions and infancy problems, especially with the injector (a further argument for Injector II), 62 microamperes of beam on target were reached. The average beam intensity was 10 to 15 microamperes, but the beam channel had been designed for 100 microamperes.

The physicists were calculating an unavoidable loss of 10% upon beam extraction. To counteract such activation, the proton channel was designed without synthetics and contained only metal gaskets. In addition, it was designed for quick and easy replacement of components, without having to overburden personnel. To reduce such burdens even further, Charles Perret had built a manipulator. He christened it "Monitor" in the style of the manipulator in Los Alamos, which was called "Merrimac" – Merrimac and Monitor were the names of ironclad vessels in use during the American Civil War, the first one belonging to the Confederates and the last one to the Unionists. However, the SIN physicists were referring to it as "Minimac" and, unlike their Los Alamos colleagues, used it only occasionally when rebuilding the proton channel.

Following the compromise made with the nuclear physicists in the acquisition of the injector cyclotron, a 4-week operations cycle was introduced: For three weeks the injector delivered its proton beam to the ring cyclotron for experiments with pions and muons – which was called the high-energy operation. The fourth week, the Philips cyclotron was reserved exclusively for nuclear physics experiments – called the low-energy operation.

A further interruption in the high-energy operation was the use of Injector I for polarized protons. Additionally, the newly commissioned Piotron for research in medicine competed with the high-energy operation because the intended beam splitter was not yet functional. Its start-up operation turned out to be very difficult for a long time and demanded much patience from the physics researchers as well as from the physicians and their patients. However, this complicated situation was eventually solved by Claude Petitjean.

Whether it made sense to shut down the expensive ring cyclotron 25% of the time in favor of nuclear physics research, is debatable. As explained above, the situation arose because of a necessary compromise with the university institutes, and it ended with the later commissioning of Injector II.

Fig. 35: 62 microamperes reached. Celebrating in the control room in 1975 are (from left) Heiri Oehninger, Claude Petitjean, Walter Fischer, Miguel Olivo, Hans Willax (in front), Werner Joho and Kurt Zimmermann.

During the following years, the capacity of the facilities was continuously improved. Those responsible for the machinery reached over 100 microamperes on Target E for the first time. It was their goal to achieve at least 1 milliampere with Injector II, which was still in development. They were also successful in achieving the demanding goal of splitting a proton beam for the Piotron area.

A Theoretical Physicist drives the Accelerator

Walter Fischer was a distinctive personality at SIN. As a theoretical physicist, he began his work at SIN in the Theory Group. When his employment in that group was not extended, he applied to the position of shift leader in the accelerator control room. There he would develop into a top operator. Thanks to him, there was always a good beam for the experimenters because he had "a feeling" for the accelerator. Later he was entrusted with the feasibility study for the spallation neutron source which he conceptualized together with Christoph Tschalär. In addition to his good educational background, he possessed an enormously broad physical know-how. This allowed him to validate the results of impenetrably complex computer programs with simple, theoretical deliberations. Since he also had a special talent for playing "father confessor" to many colleagues and associates, he always knew what was going on with management and staff at SIN.

Well-known experts had doubted the achievement of just the 100 microamperes. Werner Joho won a bet in that connection. Cyclotron specialist Henry Blosser of the University of Michigan had predicted that the modified synchrocyclotron at CERN would reach 20 microamperes, while the SIN facility would follow with 40 microamperes. Wilfred Hirt rubbed this prognosis into Werner Joho's nose. Joho bet a dinner that SIN would always exceed the CERN machine by a factor of 10. He easily won that bet and that dinner because the synchrocyclotron at CERN never exceeded 8 microamperes.

By 1978, 100 microamperes on target were routine, and the accelerator operators were working on developing a fifth cavity, the so-called flattop cavity. It was installed into the ring cyclotron in 1979 with the goal of improving the beam. Significant was the achievement of cleanly separated beam paths all the way to extraction. Also, by means of an eccentric proton injection, the beam separation at the extraction septum could be improved. With the later substitution of the four aluminum cavities with cavities made of copper (see paragraph 6.2), the cavity voltage could be increased from 500 to 900 kilovolts. Because the particle beam needed to complete fewer rotations, better beam separation could be achieved.

Fig. 36: The SIN site in 1980. The guest house is visible at the left edge. It offered accommodations to external research groups during their experiments at SIN that lasted several weeks. New is the block of buildings attached to the experimental hall on the mountain side, containing spaces for medical research as well as for the testing of super-conductors.

Injector I was also significantly improved. Thomas Stammbach succeeded in 1978 to improve the original extraction rate from the Philips injector of 70% to 95%. This simplified service operations because the reduction of losses meant a reduction in radioactivity, and also increased the maximum current from the ring from 100 to 200 microamperes. Decisive was also the marked stability improvement of the high-frequency system achieved by Peter Sigg.

In 1980, 170 microamperes on Target E were reached; however, both targets did not match the planned higher currents. They had to be newly designed. This work fell under the high-current project (see paragraph 6.2), which required a new facility concept anyway because of the spallation neutron source.

However in 1981, experiments using the ring experienced a setback. The High-Energy operation was shut down for three and a half months due to a vacuum leak. This incident was technically investigated and required further optimization by those responsible for the accelerator. The current and beam time capacities thus could be increased during subsequent years.

In June of 1984, SIN celebrated its 10th anniversary with a reception and a symposium. On this occasion, Director Jean-Pierre Blaser stated: "There are three persons who have significantly contributed to the creation of this national center: Federal Councilor H.P. Tschudi, ETH Board President H. Pallmann, and C. Seippel, President of the Building Committee."

"...the counters were literally swooshing!"

When the accelerator operators had reached 7 microamperes, they had already exceeded the values of other meson factories. At that time, Hans Hofer was in charge of measuring pion production. Up to that time, he was used to hearing the counters tick. He persuaded Hans Willax to ramp up the current temporarily as a test. This created annoyance in the experimental hall because the people in charge of radioactivity shielding were afraid that too much radioactivity would be created and would interfere with their work. After Willax had pushed up the current, Hofer appeared in the control room and announced all excited that the counters were literally swooshing."

5 The Research Program at SIN

5.1 *Introduction*

According to its mandate, SIN made its research facilities available to scientists of Swiss universities. Since the facilities were unique in Europe, they were also much in demand among researchers from other countries.

Its research objectives can be explained with the following schematic

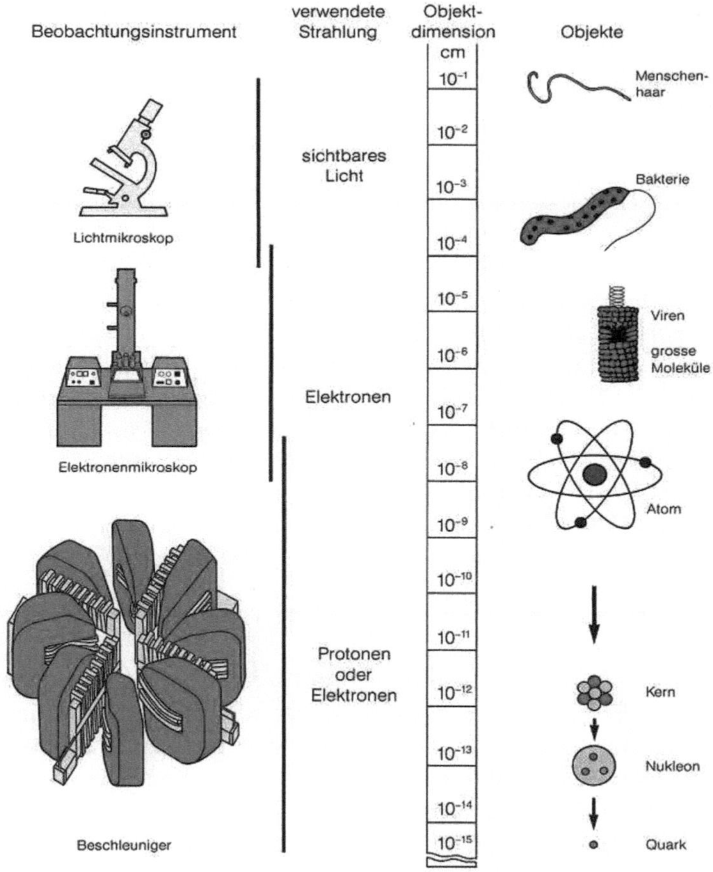

Fig. 37: Research instruments for objects of various sizes

The directors of SIN were not just interested in running a user laboratory. They wanted to see groundbreaking research and top results. To achieve this goal, they utilized two tools: First, quality control within a network of scientists of international repute; secondly, targeted deployment of institute resources in order to facilitate high-quality projects.

The will of SIN management to promote excellent research was formulated as follows in a SIN brochure (undated but probably printed in 1974): "SIN's facilities are to be adapted exactly to research requirements. Precisely analyzed particle beams, whose properties can be chosen for each experiment, can be made available to scientists at their experimental stations by means of magnetically focused channels."

SIN was prepared to make costly investments in chosen projects. In order for this largesse not to be misunderstood, management insisted on being thrifty with the available budget.

About Thrifty Treatment of Research Funds

SIN management placed a high value on thrifty use of research funds. All luxuries were frowned upon. This was applied especially to travel expenses. For attendance at conferences, Director Blaser's motto was: "Better four participants in a 2-star hotel than two participants in a 4-star hotel." Generally, SIN, with the agreement of staff, paid less for travel expenses than was required by the Confederation. This made it possible to demonstrate for the Federal Finance Control, which could not be expected to judge the necessity of expensive research instruments, that budgets were being respected. By the way, there was a rumor making the rounds among physicists that the woman handling travel expenses in the SIN-administration had tried to impose the use of a youth hostel on a researcher...

The experiments conducted at the SIN facilities were more extensive than generally assumed. They were conducted in an often international collaboration of various universities, consisted of up to ten scientists, some of them associated with SIN itself, and the projects lasted several years. A group of interested scientist could suggest an experiment. All suggestions were then reviewed by an international committee and submitted to a plenum of all users – the so-called user assembly. If a suggestion found support, SIN prepared an appropriate research area and made available beam time, shop capacity, computer services and sometimes money for equipment, such as detectors or special magnets, or for preparatory work at a university. Usually, the desired beam time exceeded available capacity, so that in the end a program committee, headed by the director of SIN, had to make the decision.

Often, one year would pass with theoretical planning and construction of research equipment before the particle beam was ever used. Then scientists would conduct their measurements at SIN over a period of time that could stretch to several weeks. This would be followed by analysis which in many cases required the use of a supercomputer.

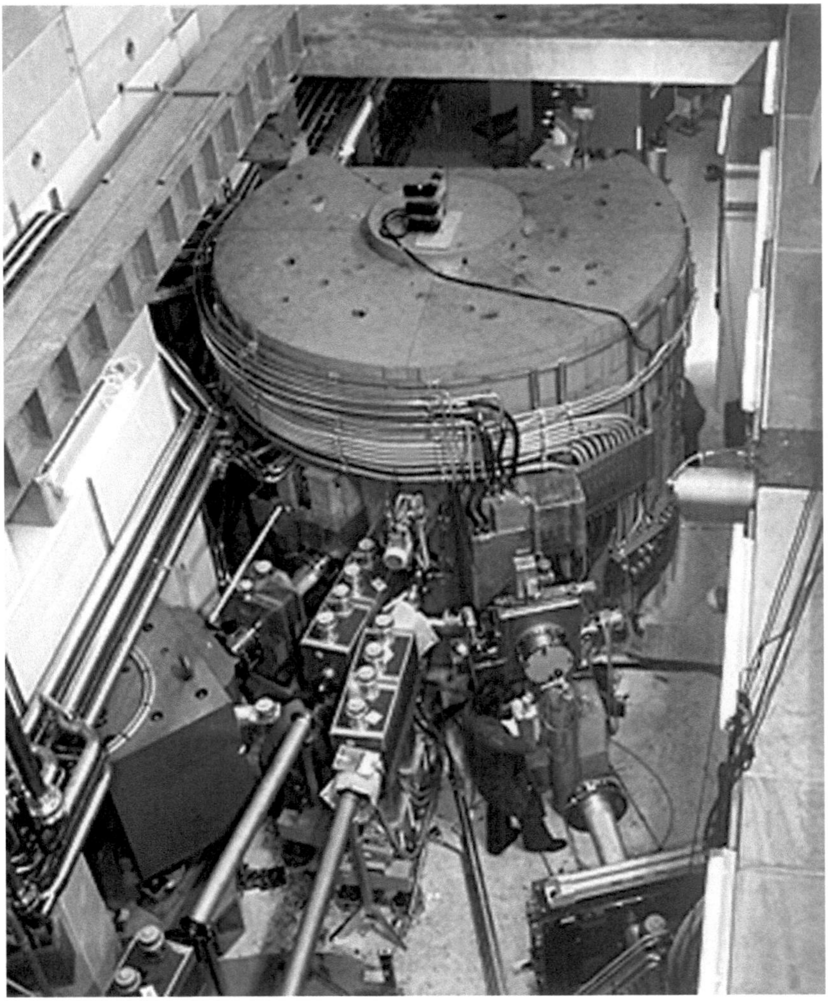

Fig. 38: Production Target M with view of the secondary beam tubes

In a typical experiment at SIN, elementary particles, such as pions or muons, were used as projectiles to be bombarded onto a piece of material, the so-called experimental target (not to be confused with the production targets which delivered pions and muons). This caused a nuclear reaction in the target which manifested itself in various ways: The projectiles were diverted ("strewn"), braked, or absorbed by nuclei, or else new particles were formed. From the number and the nature of such reactions, researchers were able to draw conclusions about the forces between the elementary particles or the properties of the target nuclei.

Fig. 39: Proton channel to Target M

The pions produced in great numbers at SIN became the primary projectiles, but also muons were used. π-mesons, pions for short, are created by a collision of accelerated protons with Production Targets M and E. There are positively charged, negatively charged and neutral pions. Muons, a type of heavy electron, are created by decay of the relatively short-lived charged pions. They have an either positive or negative charge. Other appropriate projectiles from the SIN facilities were protons, neutrons or even light nuclei, such as deuterons and alpha particles (helium-4 nuclei). There were a great many of such secondary beams, but they were either mutually exclusive or could be used simultaneously only under certain circumstances. With the subsequent development of long-term experiments, this great variety was no longer possible.

Fig. 40: Experimental area. A secondary beam is coming from above left; an experimental target is being built in front.

The scientific fields that were dealt with at SIN were many. Aside from the primary fields of elementary particle and nuclear physics, research in solid-state and chemical solutions investigated by muons and pions as probes, gained increasing importance. Experiments with negative pions on biological substances became the basis for cancer therapies on humans. Therefore, the SIN director already wrote in his 1974 annual report: "The stimulating collaboration among various natural sciences – such as nuclear and particle physics, solid-state physics, chemistry and up to biologic and medical applications – creates an exciting scientific atmosphere." The trend towards research diversification continued. The facility concept of SIN was so flexible that it allowed the incorporation of new subjects after old research directions had run their course.

On the other hand, the upgrading of the SIN facilities towards ever higher beam currents (see the following chapter) significantly influenced the research program in that the higher currents permitted even more sensitive and exact measurements.

Also, as spin-offs of the accelerator facility development, additional projects in applied sciences presented themselves over time and gained importance in SIN's research program.

The annual report of 1985 points to the ambitious experimental program. During this typical year, SIN documented 100 experiments with 80 participating institutes. SIN scientists participated in many of the experiments, and also had primary responsibility for the experimental equipment. Later on, the institute strengthened its own research. To provide a suitable framework, it created the research department SINFA under the leadership of John Domingo.

In subsequent years, the trend towards ever more refined experiments continued, thanks to the high-current beam. This created increasing demands on computing power, i.e., the processing of data gained during experimentation. As a result, this activity takes up ever increasing space in SIN's annual reports beginning in the mid-1980s.

After SIN became PSI, the importance of particle and nuclear physics decreased overall because of the new research directions that used the accelerator facilities.

5.2 *Particle Physics*

This field of science classifies elementary particles in accordance with their electrical charge, mass, magnetic power and other characteristisc, as well as the structure of interactions or forces to which they are subject. Ignoring gravitation, which does not play a role in particles of such low mass, the researchers at SIN worked on all other known interactions, i.e., electromagnetic, weak and strong. All charged particles are subject to electromagnetic forces. Muons and electrons are also subject to weak powers. Nuclear force concerns nucleons (protons, neutrons) as well as pions. Research in particle physics attempts to attribute the forces and the types of particles to a few basic elements. That the electromagnetic and the weak interaction are manifestations of an electroweak force was discovered in the 1970s. In order to also include strong interactions, SIN's researchers devised several experiments.

Some of them concerned the question whether neutrinos, which are created by the decay of nuclei, pions and muons, have a mass other than zero. Experimenters were successful in again significantly reducing the heretofore upper limit of neutrino mass. Another experiment, led by Nobel Prize winner Rudolf Mössbauer, which was conducted at the nuclear power facility Gösgen, dealt with the possible transitions between several neutrinos, i.e., neutrino oscillations. The highly sensitive experiments showed no oscillations.

Fig. 41: The approximately 8 meter high SUSI (SIN Universal Spectrometer Installation) pion spectrometer

Generally, the high-intensity proton beam extracted from the ring permitted experiments with nuclei, pions, muons and neutrons in order to examine actually forbidden or very seldom-occurring events and to conduct precision measurements which determined the upper limit for the occurrence of an event.

Fig. 42: The experimental hall at SIN. The proton channel is to the right underneath the concrete shielding blocks. To its left, experimental areas for secondary beams on Target M.

The examination of the electroweak interactions in particle physics served measurements and experiments in the decay of pions and muons as well as the capture of muons. The strong interactions were examined using pion-nucleon and nucleon-nucleon scattering experiments.

Beginning in 1981, ever larger detectors were developed for experiments at the ring which utilized the top capabilities of electronics to examine central questions of particle physics, such as the conservation of the number of leptons in the decay of muons in three electrons. Important pieces of equipment in this connection were the enormous SUSI pion spectrometer of 1974 (fig. 41), the LEPS (a spectrometer of the University of Karlsruhe for the interactions of low-energy pions with nuclei), and the SINDRUM (for seldom-occurring muon decays), which all were commissioned in 1983. In 1985, the BGO calorimeter followed for rare pion processes. A detailed description of these detectors and their measuring capabilities would go beyond the intent of this story. In this regard we reference the document written by Peter Truöl.

In summary, it can be stated that the very high intensity of the pion and muon beams permitted the study of rarely occurring decay and to reduce the limits of such decay by an order of magnitude. Rarely occurring decays are actually prohibited because of their assumed conservation laws; at least they cannot be explained under the current standard model. The search for rare or prohibited processes was the most important branch of particle physics at SIN, and it had the largest influence on science. As Peter Truöl wrote in his summary description, "here, medium-energy physics can absolutely compete with highest-energy physics."

5.3 *Nuclear Physics*

The primary accent of nuclear physics conducted at SIN lay on the reaction of pions and muons with nuclei in order to investigate the structure of nuclei. The physicists researched atoms with pions and muons, where the nucleus caught the corresponding particles, as well as muonium (an exotic atom containing an anti-muon and an electron); the scattering of pions, muons and nucleons by atomic nuclei; the bombardment of nuclei with polarized protons; the scattering of polarized deuterons by nuclei of helium.

Most of the nuclear physics was conducted with beams from the Philips injector. That was the reason why this injector continued to be used, even after Injector II had been commissioned.

5.4 *Applications*

Muons and pions could be used not only for experiments in particle and nuclear physics, but also for analysis of chemical structures in atomic and solid-state physics. Especially positively charged muons can be inserted as probes into solid-state materials and chemical solutions. If polarized muons are stopped and then decay in a material, one can draw conclusions about the local magnetic field by measuring several parameters which were "witnessed" by the muon while it was alive.

Muons in hydrogen can bind two nuclei into a type of molecule where they are so close together that they cause a nuclear fusion. That a single muon could cause more than one fusion was something that was proven for the first time at SIN.

Pions found application in biology and – importantly – in nuclear medicine. For this purpose, a pion beam that was aimed downward had been developed early on in

Pion Area 3 of Target E. Another application, which would be expanded into a larger program, was the production of isotopes for applications in medicine.

The medical programs as well as additional important applications – super-conductivity, nuclear fusion, and neutron source – are discussed in chapter 9.

6 Upgrading the Facilities

6.1 *The Injector II*

When the injector cyclotron was originally chosen, SIN's founders had decided on the Philips cyclotron for reasons of science policy, knowing that this choice would limit the entire facility to a beam current of about 100 microamperes. Once the injector was commissioned, the accelerator crew was able to continuously increase the current and the quality of the beam. By 1977, the improvement possibilities of the accelerator facility had been exhausted due to Injector I.

However, a high-intensity proton beam had been the goal of SIN from the beginning. Therefore, SIN management had decided early on to conceptualize a second injector which would make a higher performance of the ring accelerator possible. With the new machine, the goal was the extraction of a proton beam from the ring of at least 1 milliampere. This would necessitate investments in the entire facility, but would open up new research possibilities.

In addition to achieving a higher intensity beam, another argument in favor of the construction of Injector II was the fact that the ring cyclotron was shut down every fourth week because Injector I was reserved for nuclear physics experiments. This absurd circumstance was illustrated drastically by Werner Joho during a lecture with an analogous example: Swiss Formula 1 driver Regazzoni had to skip every fourth race because his wife needed the race car for going shopping. Never, said Joho, had he achieved more laughs in a lecture than with this example.

The concept for Injector II was finished in December 1973. SIN had named a project management team consisting of Hans Willax, Urs Schryber, Werner Joho and Miguel Olivo. With the creation of the new "Applications/Projects" department, which was headed by Urs Schryber and was developing Injector II, Hans Willax left the project management team. The time line called for start of construction in 1974 so that beam testing could begin end of 1977. The project was delayed, primarily for budgetary reasons.

Since internal financing, as originally planned by Blaser, did not work, a credit of 15.3 million francs for SIN Injector II was requested by the Federal Council and presented to the Swiss Parliament with the ETH building budget proposal of 1977, and it passed without problems.

Until the planned commissioning of the new machine in 1982, this new injector was the top project at SIN. SIN's accelerator builders proved to be rather courageous by designing the Injector II already back in 1972 at a time, when the functionality of the ring cyclotron still had to be proven. And for the suppliers of the Philips injector this was especially irritating!

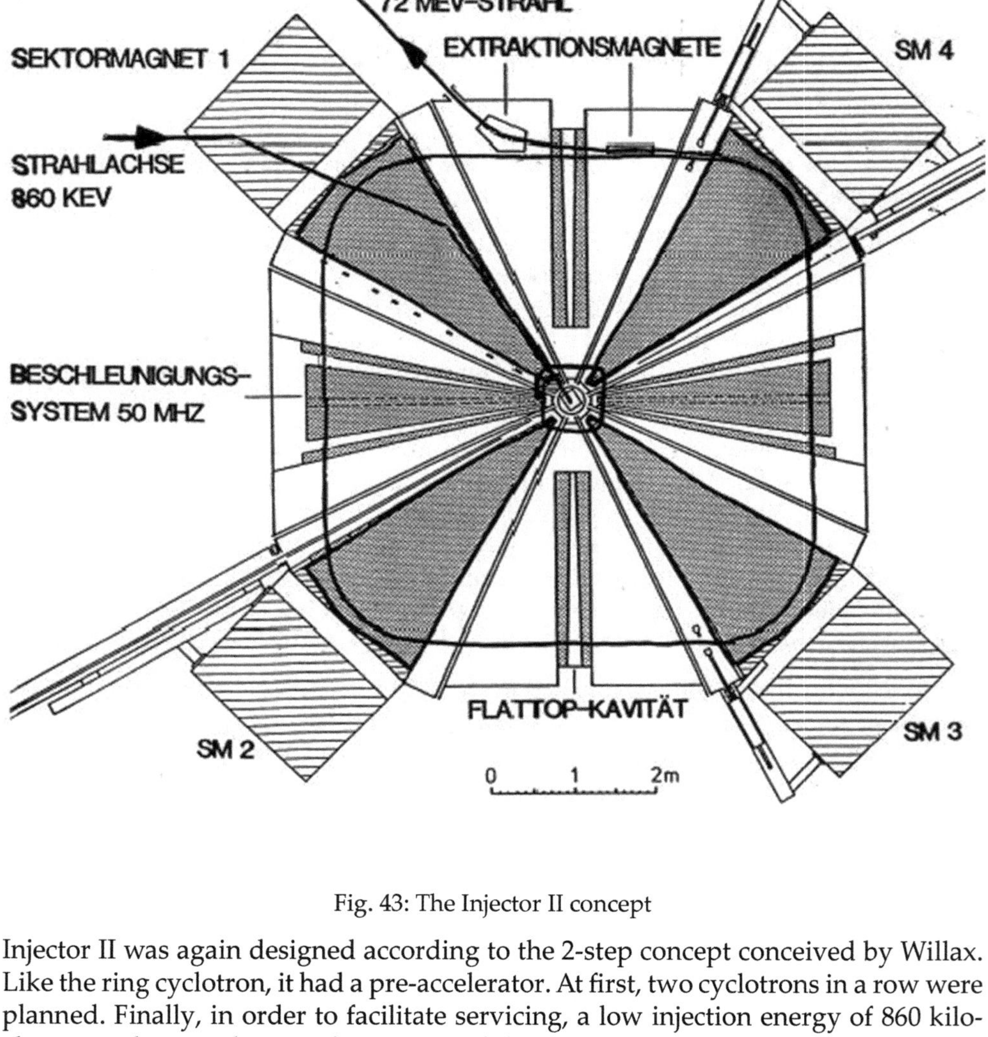

Fig. 43: The Injector II concept

Injector II was again designed according to the 2-step concept conceived by Willax. Like the ring cyclotron, it had a pre-accelerator. At first, two cyclotrons in a row were planned. Finally, in order to facilitate servicing, a low injection energy of 860 kilo-electron volts was chosen. This permitted the use of an electrostatic pre-accelerator. Because of limited space for the injection into the Injector II Cyclotron, this had to have a very large radius. It turned out that by this enlargement the weight of the magnets did not increase because the magnetic field became weaker. The pre-accelerator consisted of a commercially produced Cockroft-Walton high-voltage generator of 860 kilovolts and a proton source developed by SIN. The protons were fed into Injector II, where they were accelerated to 72 MeV and guided into the ring.

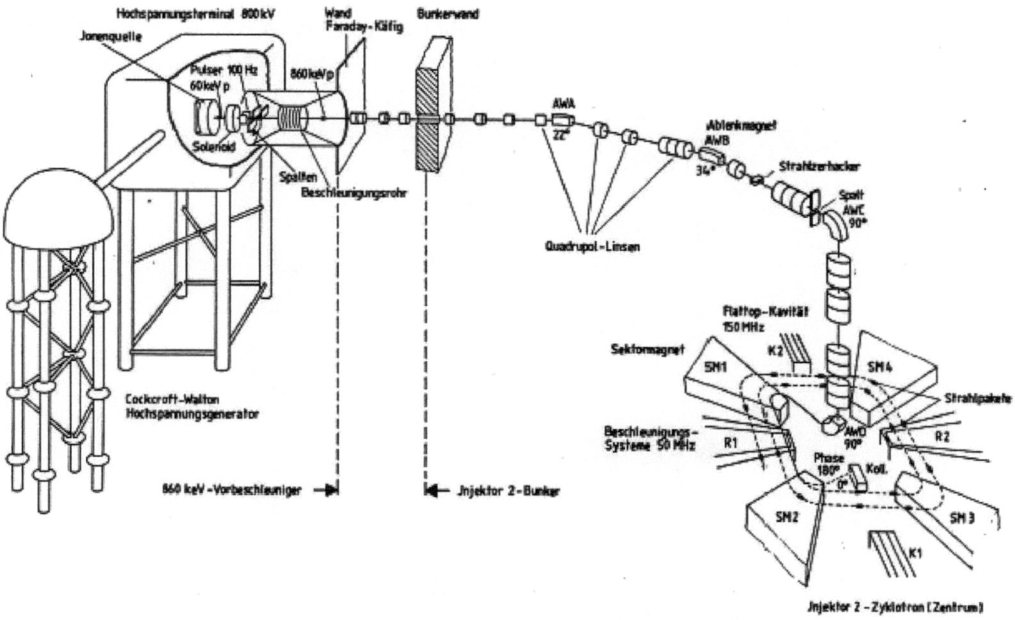

Fig. 44: Schematic of feeding Injector II. The Cockroft-Walton high-voltage generator is to the left, and above it is the high-voltage terminal with 860 kilovolts and containing the ion source. After leaving the acceleration tube ("Beschleunigungsrohr"), the protons are guided by diversion magnets into the Injector II bunker and then fed into the center of the injector from above.

Stefan Adam, who participated significantly in the design of the ring's sector magnets, demonstrated through his beam calculations for Injector II that the originally longish beam packet deformed at high currents into a small sphere due to space charge forces. It was through this effect that it became possible to extract beam currents of 2.7 milli-amperes from the injector.

During the anniversary year 1984 the pre-accelerator had been tested successfully, and Injector II was commissioned. However, it could deliver only 50 microamperes into the ring because the targets for the pion production could not accept proton beams above 170 microamperes. They had to be rebuilt, and that also went for the ring accelerator as well as the entire beam facility in the experimental hall. This led to the next large project, the so-called high-current upgrading.

At the recommendation of the Swiss Science Council, SIN management decided to continue the operation of Injector I because of experimental programs in so-called low-energy research that were still attractive. Researchers involved in these programs now had the advantage of full use of the machine because it was no longer needed as injector for the ring. This research activity put heavy emphasis on experiments with polarized particles. An appropriate neutron beam was installed for the first time. The Philips cyclotron also saw more use for the production of isotopes and the eye cancer therapy with OPTIS.

Fig. 45: Cockroft-Walton High-Voltage Generator for 860 kilovolts

After commissioning Injector II, the accelerator facilities at SIN could be characterized according to the following configurations:

The SIN Accelerator Facilities

Injector Cyclotron I (Philips)

Isochronal cyclotron with 4 spiral sectors
490-ton electromagnet
Power consumption approx. 1 megawatt
Delivers
Protons of 10 to 72 MeV
Deuterons of 10 to 65 MeV
Helium-4 nuclei of 20 to 130 MeV
Is being used since 1985 for low-energy physics and medical applications

Injector Cyclotron II (SIN), commissioned in 1984:

Isochronal ring cyclotron with pre-accelerator
4 sector magnets and 4 high-frequency resonators
Pre-accelerator: Cockroft-Walton, 860 keV
Total weight of magnets 720 tons
Power consumption 1.2 megawatts
Delivers up to 2 milliamperes protons of 72 MeV

Ring Cyclotron (SIN):

Isochronal cyclotron with 8 spatially separated magnets (spiral sectors) and 4 high-frequency resonators inserted between the magnets
Total weight of all magnets 1944 tons
Total diameter 15 meters
Top resonator voltage 600 kilovolts
Power consumption approx. 2 megawatts
Delivers currently up to 190 microamperes of 590 MeV protons

High-energy particle beams with approximately 160 beam guidance magnets

Radiotion shielding:

27,000 tons of normal concrete blocks
7,400 tons of heavyweight (iron shot) concrete blocks
14,000 tons of iron blocks

Current experiments (1984):

10 Low-Energy
45 High-Energy
10 Biology

Fig. 46: Injector II

6.2 *The High-Current Project*

The commissioning of Injector II created the basis for the upgrading of the facilities to beams of several milliamperes. At 600 MeV, this meant that the proton beam reached a power range of more than 1 megawatt – not a trivial size considering the necessary radioactivity shielding and the safety of the beam guidance system. Limited financing for SIN, brought about by Federal and ETH-domain budget reductions, caused management to postpone by several years the originally planned quick upgrading of the ring accelerator, beam guidance system and secondary beam targets, and to plan for an intermediate step of 0.5 milliamperes.

Before the high-current beam delivered by Injector II could be accelerated in the ring cyclotron, its high-frequency system needed to be expanded. Load problems caused in the closed loop by the proton beam had to be resolved first. Only after that had been completed, could the new construction of the four high-frequency power stages (of 1 megawatt each) as well as the beam extraction system and radioactivity shielding follow. The cavities were also newly developed and replaced in 1985. The expansion of the high-frequency system, which contributed enormously to capacity increases, was mastered brilliantly by department head Urs Schryber, together with the high-frequency team under Peter Sigg, after Schryber had been successful in persuading management to release the significant financial resources.

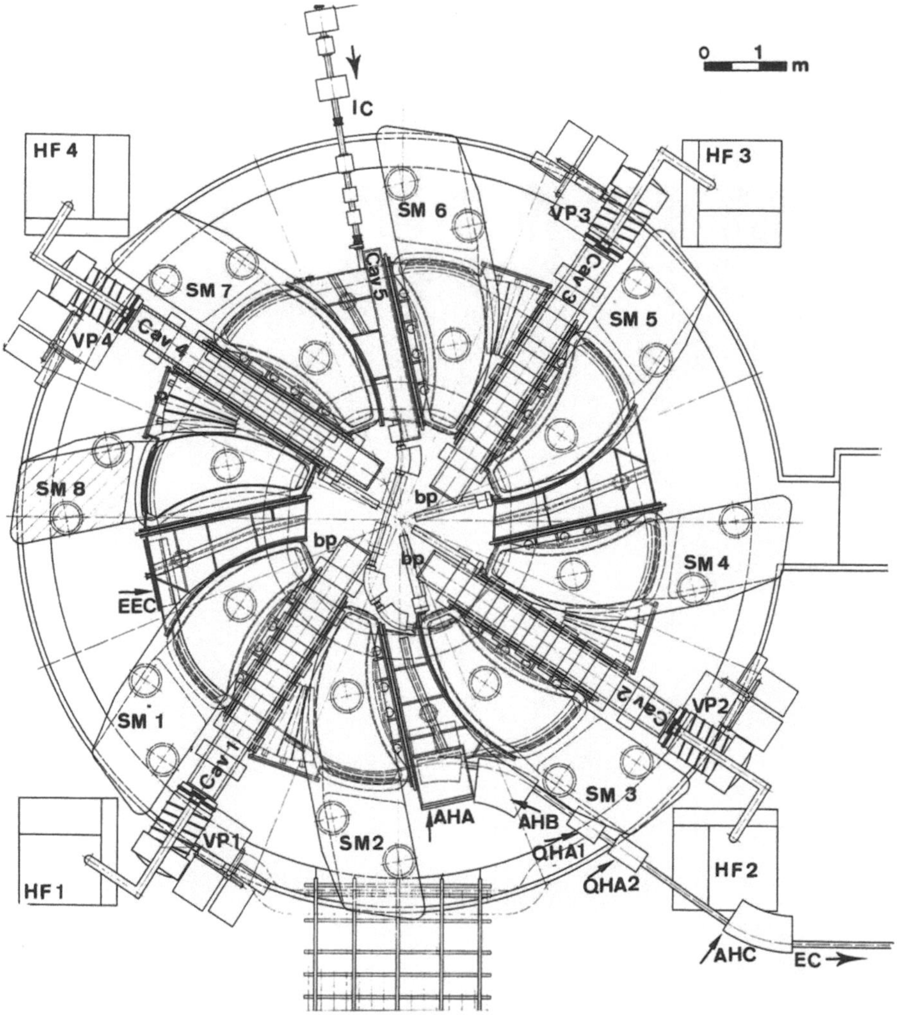

Fig. 47: Ring accelerator 1979: IC = injection canal coming from Injector II, SM = 8 sector magnets, Cav = 5 cavities (incl. flat top), HF = output stage of the high-frequency generators, VP = vacuum pumps, EEC = extraction canal of the 590 MeV protons, AH = diversion (extraction) magnets, QHA = quadrupole magnets

The high-current project had been budgeted at a total of 24 million francs, not including the new copper cavities which cost an additional 20 million francs.

The first step towards high-current expansion was taken in 1985. Project leader for beam guidance and targets was Erich Steiner. Target M was successfully rebuilt. The old station was removed and a new, pre-assembled one put in its place. This was a challenging task because, after 10 years of operation, many components had become so radioactive that only sophisticated work methods could lead to success. Now it was possible to extract 1 milliampere from Injector II and to bombard Target E with 300 microamperes.

Fig. 48: Dismantling Target Station M, 1985. Personnel in radiation protection suits.

Fig. 49: Installation of the largely pre-assembled Target Station M

Additional upgrading work included the Target E (planned for 1988), the secondary beams, the beam dump as well as, finally, the spallation neutron source. Additionally, new and sometimes costly experiments that had been made possible by the high-current expansion (see paragraph 5.2) necessitated the lengthening of the experimental hall by 50%. Also in the planning stage was the construction of a proton beam with low intensity adjacent to the experimental hall, which would permit experiments with polarized protons – and later neutrons – on the one hand. On the other hand, it was intended to serve PIREX, an experiment in radiation damage (see paragraph 9.4), and make a facility for cancer therapy possible.

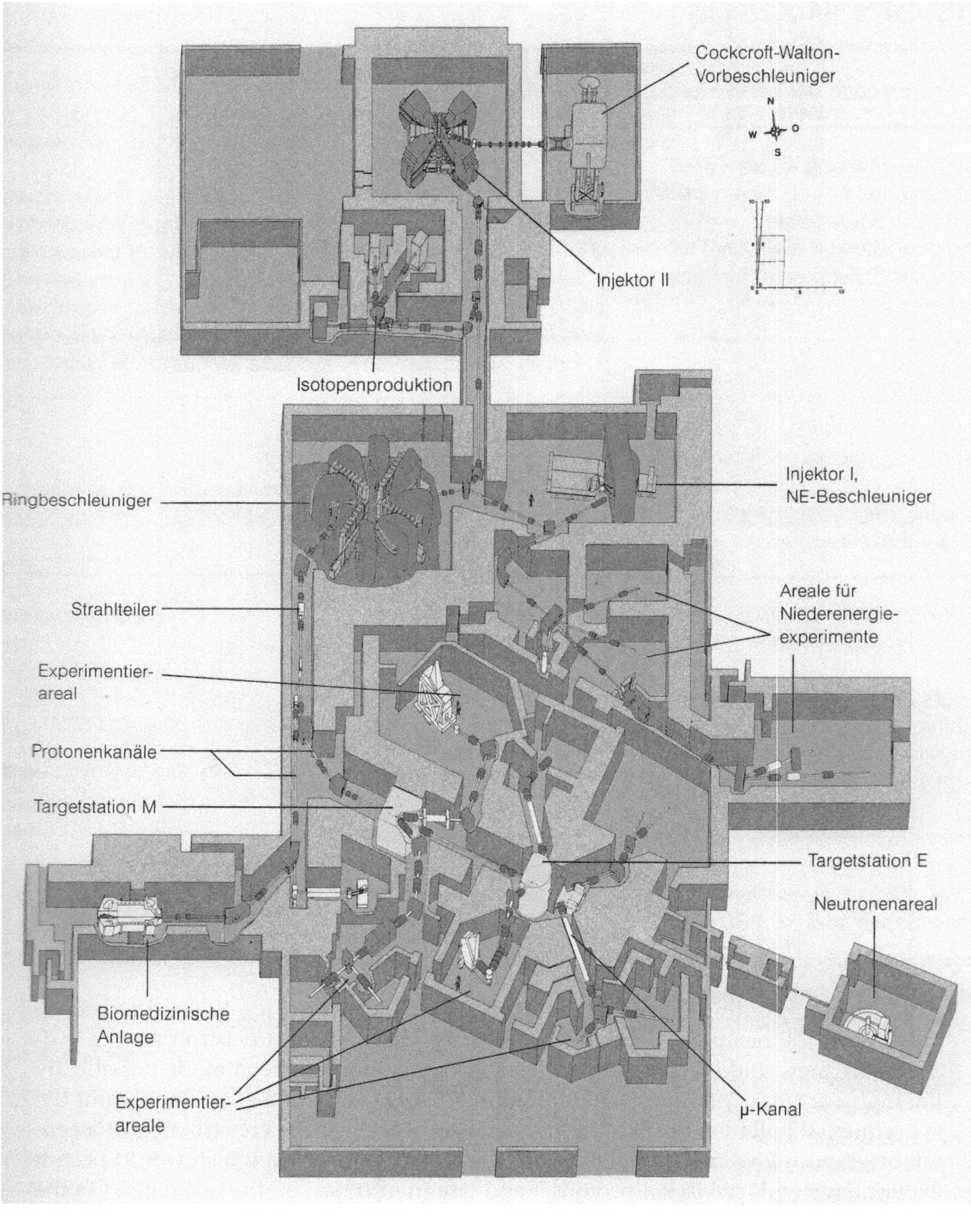

Fig. 50: Layout of experimental hall 1985

All of these measures were taken prior to SIN becoming PSI but, for the most part, they would not be realized until under the aegis of PSI. The layout of the experimental hall, as it was planned in 1982 for 1986, can be seen in photo 50. Because of federal budget cuts, the extension of the experimental hall was delayed. Nevertheless, preparatory work could be completed in 1985 to the point where in 1986 construction of the foundation and the foundation slab could begin. Occupation was planned for the spring of 1987.

Fig. 51: The SIN site in 1984 with the new central building (left)

Milestones of the SIN Acceleration Facility

1939	*Commissioning of the ETH Cyclotron under Paul Scherrer*
1958	*Paul Scherrer founds the Cyclotron Planning Group at ETH*
1962	*Jean-Pierre Blaser, the successor of Scherrer, sets a new goal of a "meson factory" with 500 MeV protons and a beam current of 100 microamperes. Hans Willax launches the idea of a ring cyclotron.*
1964	*International expert express strong doubts about the ambitious goals. Domestic scientists are opposed because of costs. However, the project is supported by Federal*

Councilor Tschudi, ETH Board President Pallmann as well as by the Swiss Science Council.

1965 *The Swiss Parliament approves a budget of 92.5 million francs for the project.*

1965 ff *Testing of 1:1 prototypes of the sector magnets and the high-frequency cavities begins at the MFO Oerlikon facility.*

1968 *Foundation of SIN as an annex institute to ETH.*
Agreement with Philips to construct an injector cyclotron (Injector I).

1969 *Construction begins in Villigen*

1972 *Preliminary planning for Injector II. Swiss Parliament approves an additional budget of 34.45 million francs because of inflationary cost increases.*

1973 *At SIN's open house, approximately 2,500 visitors visit the completed ring cyclotron. In August, Injector I delivers protons for the first time.*

1974 *On January 18, 590 MeV protons are extracted from the ring cyclotron, on February 24, the first pions are created at the target.*
Experiments with pions and muons begin in the summer. The official dedication of SIN takes place in October.

1975 *The super-conductive muon channel is commissioned and produces the most intensive muon rays worldwide.*

1976 *On December 21, the original goal of 100 microamperes from the ring cyclotron is achieved.*

1978 *The idea of a spallation neutron source is presented at the International Cyclotron Conference in Indiana.*

1984 *Injector II is commissioned.*

1986 *A budget of 33 million francs is approved for the spallation neutron source.*

7. The Cooperation with Universities

Conceived as a user laboratory for experimenters from Swiss as well as foreign universities, SIN made beam time as well as technical and computer infrastructure available to them for their research. For this purpose, a detailed user guide was issued in English which described the range of services. To complete their experiments, researchers would typically stay at SIN for three weeks. Since appropriate accommodations were not available in the area, SIN, together with interested cantonal universities, created a foundation which built and operated a guest house. It opened in 1974 and was enlarged in the 1980s. For guest scientists, who stayed for extended periods with their families, SIN rented a few guest apartments in Brugg-Lauffohr beginning in 1975.

Fig. 52: The industrial-looking, unpretentious reception office of SIN

The first area of contact for guests was SIN's reception office. From here, Christine Bertsch tirelessly, and sometimes at the last moment, organized travel for scientists of SIN as well as guest scientists and managed to not just meet travelers' expectations, but to also find the most advantageous deals. Mrs. Bertsch was helpful but not subservient and, on occasion, would make a fitting remark to a frantic researcher, such as "is there a fire under your rump, doctor?"

Aside from the Building Committee, which accompanied the construction of SIN, the ETH Board also created a Science Committee which advised the director in matters of scientific research. The following professors were members of the committee: Beat Hahn (Bern), Kurt Alder (Basel), Gerhard Backenstoss (Basel), Eugen Baumgartner

(Basel), Anselm Citron (Karlsruhe), Hedi Fritz-Niggli (Zurich), Walter Hälg (ETH), Roger Hess (Geneva), Wolfgang Horst (Zurich), Pierre Huguenin (Neuenburg), Willy Lindt (Freiburg), Jean Rossel (Neuenburg), Christoph Schmid (ETH), Jean-Pierre Schneeberger (EPUL/EPFL), Hans H. Staub (Zurich), Gerard Wanders (Lausanne). Professor Citron came from the Nuclear Research Center in Karlsruhe. The other members represented various fields of expertise at Swiss universities. This committee co-created the interface between SIN and the universities.

The close cooperation with universities was documented by SIN by the participation of professors from Swiss and international universities in all its committees and advisory commissions. As documented in their annual reports, the committee structure was highly developed and in accordance with the widely varied programs at SIN.

Beginning in 1970, even prior to commissioning the SIN facilities, SIN also conducted conferences in order to discover the needs of users. Also, to promote an efficient research operation, SIN conducted various courses, such as a summer school in medium-energy physics in 1970 in Leysin, and after that, regular summer schools in Zuoz, each of which was dedicated to a specific scientific field. Already in 1971, SIN conducted a symposium for applications of the SIN accelerator in medicine and biology, and additionally the "Bürgenstock Meeting" regarding solid-state research and chemistry with muons. SIN had invited international experts in specific fields of expertise both times. By conducting these events, the institute influenced science policy in that it initiated new fields of research in Switzerland that were made possible by its facilities. And by the way, the Bürgenstock Meeting had lasting effects. One of its participants, Alex Müller, who was later awarded a Nobel Prize, repeatedly pointed out in the 1990s that a similar meeting should be organized for the planned SLS Project of PSI. That suggestion was realized in Giessbach in 1995.

With the beginning of operations in 1974, when experimenters were able to start their research with particles available at SIN (especially pions, protons, muons, neutrons), SIN management created the institution of User Assembly in which all interested researchers were represented. It consisted of one panel each for experiments on the ring cyclotron, the Injector I, and applications in biology and medicine. Remarkable was the fact that even at the concept development stage for the facilities, the research spectrum clearly went beyond particle physics. Purpose of this institution was the shaping of opinions regarding submitted proposals. A small committee of users judged the proposals, set priorities and made recommendations to the program committee as to which proposals should be given beam time and other resources. SIN Director Jean-Pierre Blaser was chairman of the program committee. In the final analysis, SIN management made the program decisions and resource commitments according to its mandate.

With the commissioning of Injector II in 1984, the User Assembly was restructured.

There now was a united User Assembly with two sub-groups "Cyclotron and Low-Energy" and "Solid-State Materials and Chemistry". For goal setting of SIN overall, management named a goal-setting committee the same year. This committee seems to have achieved little effect, which might be explained by the fact that the ETH Board, together with SIN management, had begun to think about the future of SIN and EIR.

Fig. 53: The central building with auditorium and library, which was dedicated in 1984, became the meeting place for researchers and SIN personnel. A work of art by Basel artist Beni Schweizer hangs in its lobby.

The closest cooperation with universities was, of course, with the ETH in Zurich. In the early 1970s, ETH and SIN agreed that the ETH laboratory for high-energy physics (LHE) should be located at SIN. This was consistent with the idea that SIN would assume the role of a basic laboratory for high-energy experiments at international centers, such as CERN. The cooperation was formalized in a 1973 agreement between SIN and LHE. In the 1980s, particle physics at ETH was restructured. The high-energy physicists were now located at ETH, while at SIN remained the Institute for Medium-Energy Physics (IMP).

When the spallation neutron source (SINQ) entered its planning phase, the laboratory for neutron scattering located at EIR became one of the most important partners in this SIN project.

ETH professors held managing functions at SIN from the beginning. Jean-Pierre Blaser, founder and director of SIN, lectured on basics of physics as tenured professor of the ETH until his retirement. With that position he was fully integrated into and recognized by the faculty. Hans-Jürg Gerber, professor at the Laboratory for High-Energy Physics, headed the department of Experimental Facilities of SIN for several years. This department was responsible for the external researchers, provided them with space and partially financed their experiments. Two of its members, Reinhard Frosch and Claude Petitjean, conducted research themselves. Hans-Jürg Gerber also had significant involvement in the decisions concerning the research program.

One institution of importance, even prior to the founding of PSI, was the SIN professorship. On the occasion of the retirement of Professor Hedi Fritz-Niggli, the University at Zurich decided to close the Institute for Radiobiology. This institute was unique in Switzerland and important to both the authorities for radioprotection as well as the activities of the EIR in radiochemistry and radiation protection. To SIN it seemed impermissible to relinquish this important field. It declared its willingness to finance the professorship at the university and to collaborate with the institute. The university agreed. That's how the first SIN professorship in the field of medicine got started at the University in Zurich. This model was continued by PSI with professorships for radiochemistry, among others.

8. International Collaboration and Events

In addition to SIN, there were two other meson factories in the world: LAMPF in Los Alamos (USA) and TRIUMF in Vancouver (Canada). The collaboration with both sister institutes was good from the beginning. Although they were rivals when it came to records in beam intensity and availability of facilities, the institutes had respect for one another and associated as equals. When a question arose regarding how to solve a specific problem, a sister institute readily provided information. In particular, all supported the exchange of machinery specialists. That the ancestors of the fourth director of TRIUMF, Eric Vogt, originally came from Villigen, was of no more than anecdotal importance.

The collaboration with CERN was also close from the beginning. Several key people, who later contributed to SIN's success, wrote their dissertation at CERN or had worked there as scientists. For solving the many problems that arose during the construction of SIN, CERN functioned as teacher, model or example, but sometimes also as "big brother". Although goals were being pursued at CERN that differed from SIN's, both institutes had the intention of making their facilities available to many user groups for high-quality research projects.

Where SIN profited from CERN was in solving problems of a technical, administrative and scientific nature. The form of collaboration differed from case to case. CERN specialists advised their SIN colleagues and made documents available to them. Especially valuable was the advice concerning the planning and the costs for the experimental hall and laboratory building. The participation of SIN personnel in radiation shielding and activation experiments proved to have a lasting effect since these were significant problems to be solved at SIN because of its expected high beam intensity. Finally, knowledge of pion production spectra, which were measured in a CERN/University Geneva/ETH collaboration, was important for later experiments with pions and muons. By the way, CERN groups also participated in collaborations for experiments at SIN.

Since SIN felt primarily responsible for basic science and represented a culture of openness, it harbored no fears of scientists from Eastern Bloc countries. High-quality contributions were welcome from there, and cooperation developed between SIN and CERN's Eastern bloc equivalent, the Joint Institute for Nuclear Research in Dubna. This was during the regime of Soviet Russia and for political reasons difficult and not quite as intensive as with Western institutes. Since SIN was the giving party, it was able to set conditions, and was often successful in inviting scientists, who for political reasons had fallen into disfavor in the Soviet Union, such as an outstanding physicist who, anecdotally, had forgotten to emphasize, in a preface to a textbook, the great role Lenin had played in physics theory.

Alongside Switzerland, the Federal Republic of Germany was also interested in building a meson factory. However, here a linear accelerator had priority. Since Switzerland was ahead time-wise, and the estimated costs for the German machine concept kept rising, FRG decided to forego its own machine and to "buy itself in" to SIN. An agreement was concluded in 1972 which allowed German researchers uncomplicated access to SIN and lasted until the foundation of PSI. In the beginning, the FRG paid about 2.5 million francs annually, which rose later to more than 5 million francs in accordance with the number of German users. A cooperation committee was formed which set the annual contribution and ensured that the agreement was properly adhered to. The atmosphere in this committee was always friendly, although divergent interests would arise now and then. Members of the committee on the German side were Professor Anselm Citron of the University of Karlsruhe, the current undersecretary of the Federal Ministry of Research and Technology (BMFT) as well as two representatives of the administration of the Nuclear Research Institute Karlsruhe; on the Swiss side Jean-Pierre Blaser and Wilfred Hirt of SIN management, the scientific advisor for the annex institutes of the ETH Board as well as a representative from the Swiss Federal Office of Science and Research.

A similar agreement was concluded with the Austrian Academy of Sciences, with an annual contribution of 50,000 francs.

Often an institute, which is a leader in its field, is entrusted with the implementation of a prestigious symposium. Such an event is expensive to organize, but it is a testament to the institute's aura and recognition among its peers. SIN also was invited to hold such conferences. The most important ones were:

- The "7th International Conference on Cyclotrons", 1975 in Zurich with about 230 participants from 21 countries, organized by SIN with the support of ETH and CERN. The meeting was dedicated to the state of worldwide development and applications of cyclotrons. SIN was able to present its successful operation of a meson factory just one and a half year after the first beam hit a target.
- The "7th International Conference on High Energy Physics", 1977 in Zurich with about 500 participants from 23 countries. The themes were the structure of atomic nuclei, specifically how they could be researched with mesons, as well as fundamental questions about elementary particles.
- The "International Symposium of Spallation Neutron Sources", 1978 in Villigen; this was organized by SIN together with the ETH Institute for Reactor Technology which conducted the research with neutron scattering at EIR. In addition to the neutron scattering application, the symposium dealt with new neutron sources.
- The "9th International Conference on Magnet Technology", 1985 at SIN. The basis for this conference was the expert knowledge that had been built during previous years in the area of super-conductive magnets.

Fig. 54: Cyclotron Conference 1975, from left – Jean-Pierre Blaser, Hans Frei, Werner Joho and M. Stanley Livingston in front of the ring accelerator magnets.

The Never-Ending Speech by M. Stanley Livingston

For the "7th International Conference on Cyclotrons" M. Stanley Livingston had been asked to give the dinner speech during the banquet. Livingston had built the first cyclotron in 1931 under E. O. Lawrence in Berkeley and spoke about the history of the cyclotron. Unfortunately, the organizers had failed to give him a time limit for his speech. After one hour, the French delegation in the back of the hall started to sing songs. After one and a quarter hour, Livingston had to be removed from the podium almost by force. What has to be added here is that M. Stanley Livingston had been invited at the suggestion of Werner Joho and Urs Schryber. And Director Blaser had an embarrassing mishap when he had first invited a half-brother of Stanley's, Robert S. Livingston, who was working on the cyclotron at Oak Ridge National Laboratory, and then had to disinvite him again. After the conference, Joho and Schryber agreed that Blaser had probably invited the "better" speaker.

9. New Research Fields

9.1 *Introduction*

Thanks to the potential of their machines, their entrepreneurial vision and their freedom to use their own resources, SIN had the possibility to set the course for new research fields. Such freedom is an achievement of the ETH domain. Internationally it is not usually granted because government administrations prefer giving money for research projects which can be controlled by them and are often politically motivated. Sometimes such freedom is questioned in Switzerland, but it is justified because cases where it is used for personal gain rarely occur.

The following compilation shows how such developments came about and how most of them were purposefully realized in the form of large projects.

9.2 *Particle Therapy of Cancer*

Already shortly after the discovery of X-rays, they were used for radiation therapy on cancerous tumors. After the war, the newly developed linear accelerators allowed the use of electron beams which already made possible improved dose dispersion inside the tumor over X-rays. However, since electron beams – just like X-rays – are absorbed gradually be the body, thereby depositing most of their energy into the healthy tissue before reaching the tumor, heavy proton particles were tried because they have their largest biological effect at the end of their path (the phenomenon is called "Bragg Peak") and, therefore, should permit a much sharper focus of the ray dose on the area of the tumor. Additionally, research in radiation biology showed that heavy particles have a much stronger effect than electrons: Their relative biological effect is higher. That was the reason for developing the pion therapy at SIN. After deceleration, negative π-mesons in the tissue are absorbed by atomic nuclei which explode, so to speak. In the process, heavy particles, such as alpha particles, are emitted. These cause a strong radiobiological effect and, thereby, damage to the tumor.

a) The Eye Therapy OPTIS

An early discovery of the application of particle therapy at SIN was for the treatment of ocular melanomas with protons. This is a fairly rare but difficult tumor with rapid metastasizing, making the removal of the eye necessary in most cases. Melanomas of the eye were treated for the first time in 1975 in the USA with the Harvard cyclotron. The method proved successful. However, it was not ophthalmologists who became aware of this method, but Charles Perret of SIN. This young physicist was responsible for the concept and construction of radioactive shielding in the experimental hall which, considering the size of the facility, was a large project in itself. In addition, he was also responsible for the development of the radiation protection protocol at SIN. On the occasion of a conference, where the Harvard project was also mentioned, Perret was able to arouse the interest of the Lausanne ophthalmologists around Professor Leonidas Zografos of the Hôpital Jules Gonin. Zografos was to take over the medical aspects of a joint project.

The collaboration worked extremely well, and Perret built the radiation facility OPTIS (Ophthalmological Proton Therapy Installation SIN) in 1983. For this purpose he used a portion of the proton beam from the Injector Cyclotron I. It delivered protons at 72 MeV, and that is the precise energy level needed to penetrate the size of an eyeball and stop a tumor on the retina. Thanks to the high-level technological know-how at SIN, Perret was able to introduce significant innovations when building the facility, which were later adopted by Harvard.

Fig. 55: OPTIS facility. During radiation of the melanoma of the eye, the head needs to be fixed precisely.

SIN began radiation treatments on patients in 1984. This totally unique project in Europe garnered strong international recognition and led to a great number of foreign patients, especially from Italy. The Swiss Cancer League, the Federal Office of Health and the health insurance companies assisted constructively and without creating complications in solving the unusual administrative questions regarding coverage for the therapy. Revenues increased accordingly and were a welcome increase in the research budget of SIN's medical program.

Up to the end of 2010, about 5,500 patients were irradiated (since 1988 at PSI). In more than 98% of all cases, tumor growth was stopped or the tumor was eradicated entirely. In 90% of all cases, the cancer-affected eye could be saved. Based on this success, the facility was duplicated at various European clinics.

b) Cancer Therapies with the Piotron

The above-mentioned potential of pions in radiotherapy led to a project at Stanford University (USA) where the pions were to be produced with an electron-accelerator. This concept proved to be flawed and the facility was never built. However, the meson factory LAMPF in Los Alamos built a therapeutic facility with which patients from all over the world were treated with pion rays for the first time. The same happened at TRIUMF. Both were single-beam facilities.

At SIN, which started this project after LAMPF, Jean-Pierre Blaser had been planning a pion therapy right from the start. For this purpose, a secondary beam was developed in Pion Area 3 of Target E where pions were aimed downward. Since a single fixed pion beam could not generate the desired concentration to optimally treat a tumor, there were a lot of deliberations about alternative radiation concepts. In this connection, physicist Georg Vecsey, who had developed SIN's super-conductive technology to the highest level, came up with the ingenious suggestion of a "piotron". This idea consisted of concentrating a great number of pion beams from all directions onto the tumor. Technically, this was possible only with super-conductive magnets, and those had to be developed first. As so often was the case, SIN took a risky path because it entered unknown territory.

For scanning the tumor, Vecsey was planning "smoke ring scanning", a circular dose distribution where the reach of the pions is controlled by changing their energy. This method, however, had several disadvantages that were recognized by Eros Pedroni through computer calculations during the therapy development stage. Primarily, it turned out to be extremely difficult to achieve a homogenous irradiation of the tumor with the suggested method.

The group working with radiation shielding specialist Charles Perret dealt with the development of the irradiation method. Eros Pedroni, formerly a particle physicist with in-depth knowledge of technology and informatics, suggested a method of three-dimensional scanning of the tumor with a pencil beam. He called it "spot scanning". With this method, the focal points of the sixty separate pion beams were moved within the tumor in such a way that it would be evenly irradiated with full dose while optimally protecting the surrounding healthy tissue. A simple water bolus made good conformity of beam distribution and, therefore, good biological effectiveness possible with reasonable irradiation time limits between fifteen and thirty minutes. However, instead of moving the radiation system around, the patient, secured to a table, was being moved.

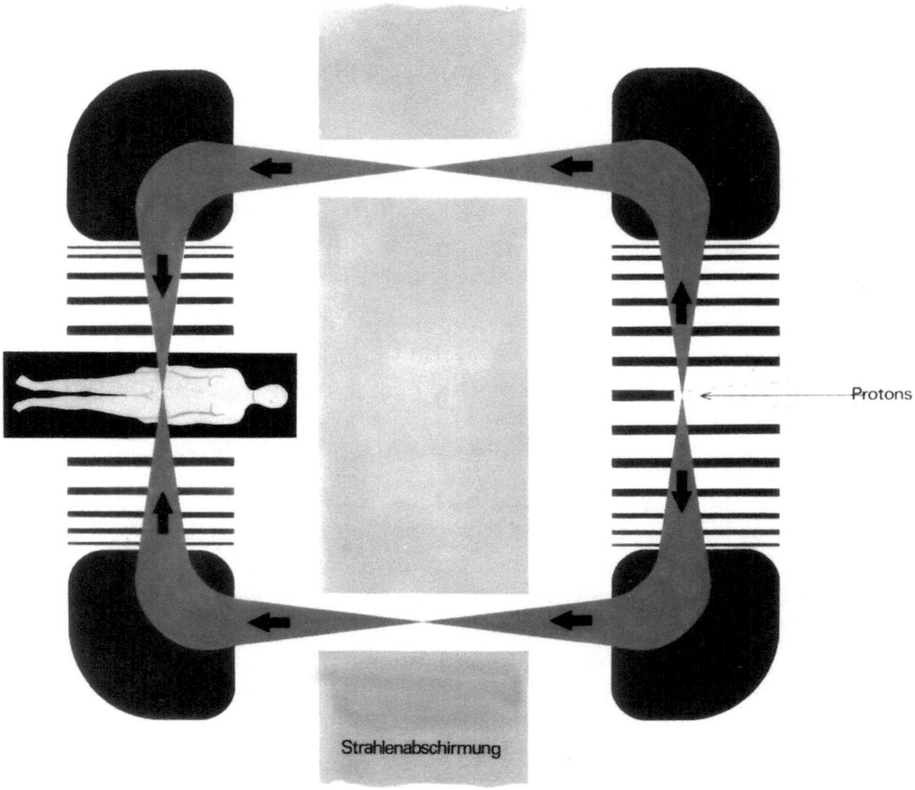

Fig. 56: Piotron schematic: Median plane with proton beam. The pions are being guided to the patient cot by means of two consecutive coils.

The subsequent realization of this method confirmed the validity of the concept, both from a technical as well as medical standpoint. Physicians were impressed with the three-dimensional spot-scan technique. It was thought to be the first system that was three-dimensional and functioned on the basis of a tumor image created with computer tomography. Fifteen years later, clinics applied it successfully by also using photons.

Unlike with OPTIS, where patients were largely mobile, the pion therapy necessitated a clinical structure. This was not a trivial matter for a large technical institute for physics research. Especially needed was a local medical director who would take the responsibility for these new clinical radiation treatments. In many discussions between SIN directors and the radiology-therapeutic clinics at Swiss university hospitals, the clinics showed interest, but no one seemed willing to take on a leading role. Therefore, the position of medical director at SIN's pion therapy project was announced internationally. Interestingly, there was not a single application from Switzerland, and none from Europe, but three from the USA, and all on the basis of a full professorship. The decision was in favor of Carl von Essen, originally from Sweden, who had great international experience in radiation therapy.

Nevertheless, Switzerland provided great support for the project. The development of a clinical operation at SIN legally required the approval of the Federal Office of Health as well as the Canton of Aargau. The Brugg hospital also offered friendly cooperation. All of this assistance was provided in an uncomplicated and un-bureaucratic way.

The support from the Swiss Cancer League was of great importance. Difficulties arose when it came to constructing a medical building for the piotron facility. Although SIN was able to finance the machine part from its research budget, it could not cover the construction of a building. That's when the Cancer League came to the rescue and endowed – all from donations – the significant sum of 2 million francs.

Now another obstacle arose: According to Swiss government regulations, the SIN Director was not formally permitted to accept the Cancer League gift. It required a resolution by the ETH Board. The then president of the Cancer League, Athos Gallino, a radiation therapist from Tessin and also mayor of Bellinzona, was invited to the ETH Board session. ETH President Ursprung – known for his astute and revealing remarks – was wondering about regulations that permitted the SIN director to spend 10 million francs of his research budget for the machine on his own authority, but had to consult the ETH Board if he could accept a gift from the Cancer League for the construction of the building.

During this session there was another interesting debate. ETH Board member Guy Waldvogel, a high-ranking director of the chemical industry, spoke out against the Cancer League gift and raised basic questions regarding the project. He thought that such expensive and large technical medical facilities were not the appropriate path towards conquering cancer. Rather, it was necessary to emphasize prevention. Since he was smoking the entire time he spoke, Gallino made the pointed remark that Waldvogel was in the middle of demonstrating how difficult prevention really is. This ended the discussion.

Construction of the piotron began in 1975. The separation of a partial proton beam to the piotron area succeeded in 1976. Considering the high power of the main proton beam, this was a challenging task. But it was important because at the time the medical facility was in competition with physics experiments. In 1977, the shell of the endowed piotron building had been constructed. Occupation occurred in 1978 and the project team was able to begin installing the facility at the operating site. Therapy planning began at the same time by the SIN medical staff, and the department "Medicine" appeared for the first time in the organizational flowchart.

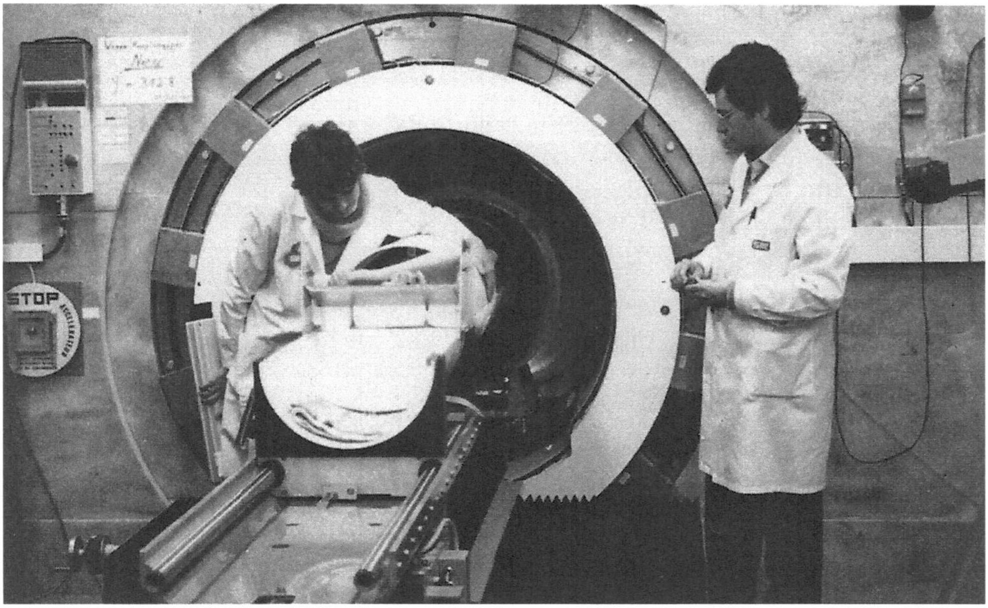

Fig. 57: Piotron. A patient is being positioned in the piotron on his individual cot.

The pion therapy project at SIN developed into an important enterprise which aroused great public interest and gained a lot of international recognition. In 1982, the start of clinical treatment of patients ran "amazingly trouble-free", as the director wrote in his annual report. First successes were shown in 1984.

A comprehensive evaluation of pion therapy could be read in a special edition of the PSI internal newsletter, Spektrum, "25 years of ring cyclotron at the Paul Scherrer Institute":

"A little more than 500 patients have been treated with the piotron by 1992. Some results promised success: primarily the treatment of large tumors in the abdomen. However, there were also reversals. After treating bladder tumors, unexpected and late complications arose several years after treatment. Also against glioblastomas (brain cancer) and tumors of the pancreas the destructive effect of pions was powerless. Looking back, pions have not turned out to be the magic pill one had wished for. The higher effect of the peak dose produced not just destruction of tumor cells, but also irreversible damage to the healthy tissue surrounding the tumor."

A leading physician at SIN, Gudrun Goitein-Kappel, had recognized that the neutrons, which are freed inside the tumor when catching the pions in the atomic nuclei, cause an additional, significant and spatially extensive distribution of the dose and influence the precision of the treatment dose. This was one of the reasons to use protons instead of pions.

Since technical problems also arose with the piotron, whose repair would have been too expensive, the pion therapy project at SIN was abandoned a few years earlier than planned.

In the final analysis, the benefit of the pion therapy project at SIN was less in extensive therapy and more for the first-time development of the spot-scanning technique. It would become the basis for subsequent projects of particle therapy at PSI and worldwide.

c) The Proton Therapy

The initiators of the pion therapy always knew that in a hospital environment, therapeutic equipment with a large and expensive accelerator, as was necessary for the production of pions, was out of the question.

Protons, on the other hand, were quite conceivable. Principally, good concentrations with spot-scanning can also be achieved with protons, but without the heavy local concentration that occurs with the absorption of pions, which can also have adverse side effects.

Therefore, SIN's next goal was the development of radiation equipment that uses protons and was based on the spot-scanning principle. Since the treatment of tumors inside the body necessitates proton energies up to 250 MeV, the project team could not use the proton beam from Injector I, as for OPTIS, but the proton beam from the ring cyclotron had to be used. This meant that the first project tests compromised the physics program at the ring cyclotron. An agreement could be reached in this regard. However, the technical challenges were significant. They could be mastered by a team headed by Eros Pedroni. Although the beam guidance concept for proton therapy at SIN had been developed back in 1984, the realization of the proton therapy equipment did not occur until the institute became PSI. This project also gained strong international significance.

9.3 PET

Since the discovery of X-rays, a goal of medicine for the diagnosis of possible illness has been to obtain images of the inside of the human body that are as precise as possible, without opening up the body. First successful tests had been conducted in the USA in the 1970s with "Positron Emission Tomography (PET)". Radioactive elements, which emit a positron upon decay (so-called beta decays of neutron-poor nuclei), are inserted into biochemical substances which then dock at desired organ sites. When such decay occurs, the positron annihilates promptly with an electron and emits two high-energy gamma rays. Their registration with detectors outside the body permits the emission site inside the body to be precisely determined.

SIN had an early interest in this method because the radioactive isotopes necessary for PET could not, as usual, be produced in reactors, but only with a particle accelerator, such as the injector cyclotron at SIN. In 1983, SIN conducted first experiments with short-lived positron emitters, and that resulted in a concrete research project in 1985.

Especially, isotopes of carbon-11 (half-life of 20 minutes), nitrogen-13 (10 minutes), oxygen-15 (2.1 minutes), as well as fluorine-18 (110 minutes) were to be manufactured at SIN. With their short half-lives, with the exception of fluorine, they had to be used immediately onsite. Remarkable in this connection was the fact that at one time it was Jean-Pierre Blaser and Pierre Marmier who had produced, with the ETH Cyclotron, the hardly-known fluorine-18 through a proton-neutron reaction of oxygen 18 and had precisely determined its half-life.

The institutes of nuclear medicine at Swiss universities, who were contacted by SIN, showed no interest in the new PET method. At that time it was unknown here in spite of having garnered placement on the title page of "Scientific American" magazine.

However, industry had heard of SIN's plans, and the then-Ciba-Geigy showed strong interest. The company recognized that PET would pave the way towards research into access and effect of medications in the brain. A cooperation agreement was concluded and the company participated with half of the facility costs. Because of lacking interest on the part of Swiss medical institutes, Klaus Leenders, a physician from Holland who had some experience with PET, was named project leader. Operations with the PET facility at SIN started in 1987.

That the first PET facility in Switzerland was created at SIN was also due to the know-how of the EIR regarding the production of biochemical substances that had been marked by radioactive isotopes. The EIR clarified the question as to which short half-life isotopes were appropriate for clinical applications.

PSI shut down the PET facility in the year 2000 because, in the meantime, several Swiss clinics had built such facilities with so-called "baby" cyclotrons.

9.4 Nuclear Fusion Technology

a) Superconductivity

Superconductivity was used at SIN early and in a pioneering fashion. First was the construction of the superconductive muon channel as well as the superconductive coils for the piotron. Thanks to this experience, Georg Vecsey's group became an attractive partner to international organizations – such as the International Energy

Agency IEA, or the EURATOM – in the development of superconductive magnets for energy storage as well as nuclear fusion.

In spite of the fact that enormous progress has been made in the 1960s in the development of superconductive technology, the construction of superconductive magnets for future fusion reactors presented an eminent challenge to the few leading research groups and industrial firms. The necessary high field force – in itself a critical parameter in superconductivity – combined with considerable dimensions – 10 meter high magnetic coils were in discussion – caused mechanical material loads whose mastery under operating temperatures of about minus 270 degrees Celsius, represented one of the most important steps towards fusion technology.

In 1977, the IEA named Georg Vecsey of SIN project leader for the construction of a Swiss coil for the project Large Coil Task (LCT) in Oak Ridge. BBC was responsible for building the coil. Altogether, six coils were developed, built and tested in a tokamak-like arrangement for the LCT project. The test facility plus three of the coils were delivered by the USA (General Dynamics-Convair, General Electric, and Westinghouse), one each of the remaining coils came from EURATOM, Japan, and Switzerland. The Swiss coil was finished in 1983 so that it could be shipped the same year to the tokamak test facility at Oak Ridge. It was tested successfully in 1986. Later it would be returned to PSI and exhibited in the experimental hall because a disposal in the USA would have been too expensive because of customs duties.

As mentioned before, superconductive magnets were subject to very high loads in a fusion reactor. Georg Vecsey's group at SIN built the superconductivity test facility SULTAN in 1979. Its intent was to facilitate the application of superconductors by using new conductive materials on the one hand and by further development of cooling technology on the other hand. An enormous cooling capacity was needed for the tests which made the construction of special equipment necessary. The test facility was ready in 1983. The 24-ton magnet system was cooled to helium temperature (minus 269 degrees Celsius) over 21 days and then tested for four weeks. The larger facility SULTAN II was in the planning stages already in 1982 and was partially financed by EURATOM and the National Energy Research Foundation NEFF. SIN made a multi-purpose building available, which was ready for operation in 1983, to house the SULTAN and the pion therapy projects.

In 1985, SIN organized the "9th International Conference on Magnet Technology" (see chapter 8). The technical know-how of Georg Vecsey's group and the equipment they had developed were of such high interest to EURATOM that it strengthened the cooperation with SIN and participated in financing several projects.

Fig. 58: LCT coil, returned from Oak Ridge to PSI. In the meantime, the research group had been transferred to EPFL, therefore the corresponding address.

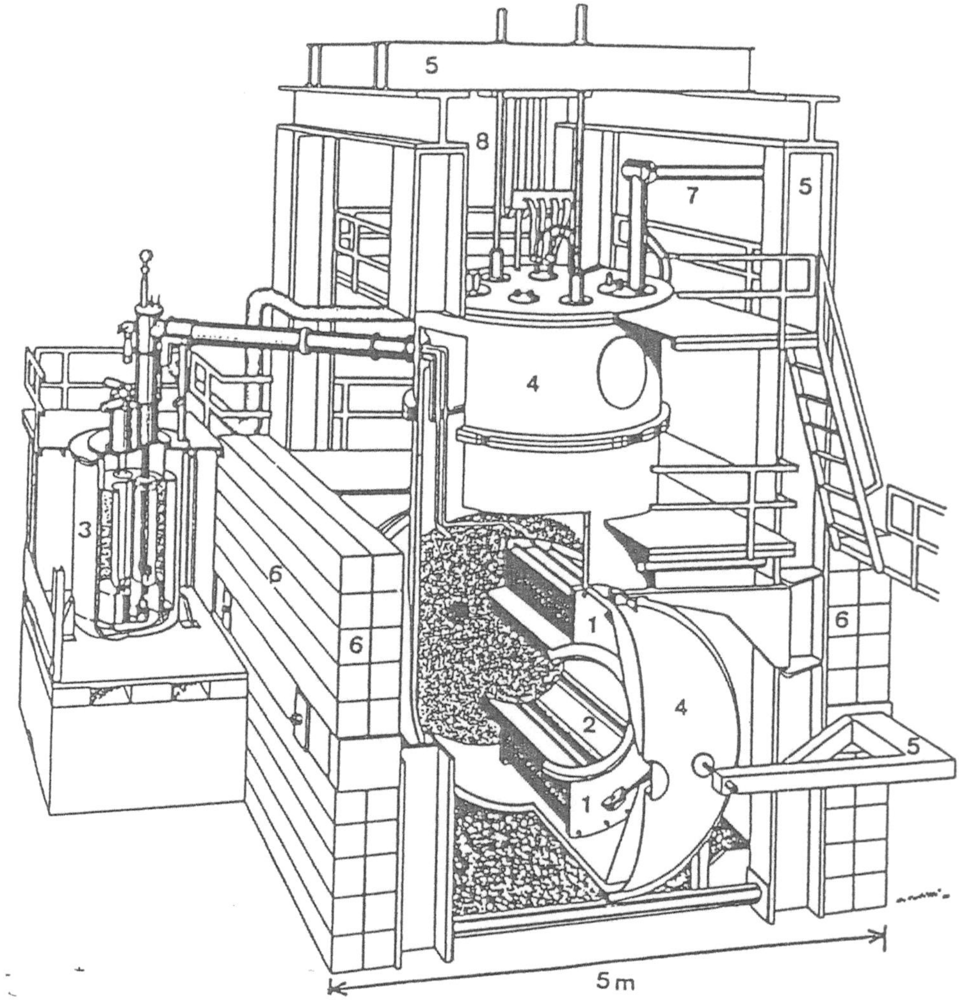

Fig. 59: SULTAN test facility: 1, 2 magnets, 3 helium cryostat, 4 vacuum tank, 5 magnet suspension, 6 radioactivity shield of iron, 7 helium line, 8 power source

b) PIREX

In the early 1980s, Max Victoria's group at the EIR discovered that the radiation damage, which would occur in fusion reactors by 14 MeV fusion neutrons, could be duplicated with the 590 MeV protons of SIN. This developed into the "Proton Irradiation Experiment (PIREX)" in 1982. It had as its goal the examination of the micro-structural material changes and the fatigue phenomena that occurred under high beam loads. Such examinations are of great importance to future fusion reactors, which is why the EURATOM participated in financing the project. Under the direction of Manfred Daum, SIN built a special area for PIREX which permitted irradiation with 20 microamperes of 590 MeV protons.

9.5 *The Spallation Neutron Source*

Since neutrons are electrically neutral and only interact with atomic nuclei (they "see" only the atomic nuclei while traversing material), they make good probes for reconnoitering the structure of materials. This had already been recognized shortly after the discovery of the neutron in the 1930s, but the first possible neutron sources were weak. This situation changed only with the construction of nuclear reactors.

The first nuclear reactors in Switzerland were created at Reaktor AG, which was founded in 1955, and became EIR in 1961. These were the research reactors SAPHIR and DIORIT. ETH Professor Walter Hälg, who had participated substantially in the construction of the DIORIT, conducted first neutron experiments on SAPHIR, together with a research group. Later on, an actual park for neutron experiments was built around DIORIT as a primary source for so-called "thermal" neutrons (the term relates to the speed of neutrons). In 1974, the ETH Board decided for financial reasons to cease operations on the DIORIT reactor. After its last nuclear fuel load was used up in 1977, it was shut down, and neutron experiments were conducted on the weaker reactor SAPHIR.

When certain materials are bombarded with protons of high speed, among other things neutrons are created by way of the so-called spallation reaction. Compared with the conventional fission neutrons in nuclear reactors, spallation neutrons offer the advantage of significantly lower thermal discharge since about 200 MeV are discharged with fission and only about one fourth of that in spallation, which is especially advantageous when producing "cold" (very slow) neutrons. Additionally, a spallation neutron source poses no nuclear-technical safety problems (less waste heat, no nuclear criticality), can be ramped down quickly, and does not require enriched uranium, which is not easy to obtain.

Because of these realities, an "Intense Neutron Generator (ING)" project was conceived in Canada in the mid 1960s with a linear accelerator of 1 GeV and a target of lead-bismuth-eutectic. However, this project was never realized.

After the DIORIT reactor was shut down, Walter Hälg, who was familiar with the Canadian project, showed great interest in producing neutrons with the ring accelerator. Director Blaser also had suggested investigating how high the proton beam current had to be in order to connect a spallation neutron source behind Target E. Hälg and his researchers came to the conclusion that a proton beam of at least 1 milliampere was necessary for the source.

Since SIN was already planning on a high-current proton beam – thanks to Injector II – in the late 1980s, it created the conditions for a spallation source. On a dark November afternoon – according to a contemporary witness – a discussion took place in which Christoph Tschalär, Walter Fischer, Erich Steiner, Jean-Pierre Blaser and others participated. The question was how 1 milliampere could be reached for the neutron source without giving up Target E. The neutron source would, therefore, use the "waste beam" after it traversed Target E.

Erich Steiner reminded everyone that Target E would drastically reduce the beam current. He suggested aiming for 1.5 milliamperes onto Target E and to shorten the target accordingly so that the pion rate would be equal to 1 milliampere achievable at the previous length. All users were satisfied and the decision was made. However, this created a great workload for Gerd Heidenreich, who was responsible for targets, until he had designed such a target.

Ideally, the energy of thermal neutrons can be determined with time-of-flight spectrometry. Therefore, Hälg was interested in a proton beam pulsed at 50 Hertz. With the SIN accelerator, possibilities existed to feed a neutron source with either a pulsed or a continuous beam. However, for a pulsed source it would have been necessary to bring the protons into an additional storage ring.

A project group consisting of Walter Fischer, Werner Joho and Christoph Tschalär of SIN and Beat Sigg of Hälg's ETH Institute for Reactor Technology conducted a concept study in 1978. During the same year, SIN organized a symposium for spallation neutron sources (see chapter 8). Discussions during this symposium resulted in the opinion that a continuous source at SIN would be the only practical solution. For one thing, the other users of the SIN facilities would not have agreed to a pulsed beam. For another, Joho was able to convince Hälg that the design of a pulsed neutron source would be as difficult as trying to get a table tennis ball into a coffee cup from a distance of five meters. Hälg, finally, was satisfied with a continuous beam for the spallation source, as long as the intensity was at least 1 milliampere.

The project was continued on this basis and was included in the new facility layout for the experimental hall in 1980. Accordingly, the neutron source would be downstream the Target E.

In 1983, Walter Fischer and Christoph Tschalär of SIN successfully examined the feasibility and efficiency of a neutron source, later called SINQ. According to a novel concept, the source was to consist of a vertical target made of molten metal (lead-bismuth-eutetic which melts already at 271 degrees Celsius) into which the proton beam is shot from below. The high-energy protons create spallation neutrons of several MeV energy inside the molten metal which are slowed down in a heavy-water moderator to so-called thermal energies, as needed for neutron scattering experiments. Additionally, a portion of the neutrons are cooled down to sub-thermal energies in a moderator made of fluid deuterium ("cold source"). The moderated neutrons are then guided through horizontal beam tubes and special guides for cold neutrons to the spectrometers in the neutron hall where experiments can be conducted. On the one hand, the concept of a vertical target made possible an optimal 360-degree utilization of neutrons emitted all around the target. On the other hand, it permitted the heat removal from the zone of the nuclear reactions partially through convection.

SIN intended to apply to the ETH Board for a credit of 32.58 million francs soon. The ETH Board decided to first obtain an expert opinion on science policy concerning a new neutron source and consulted the Swiss Science Council. The Council had already supported the project in 1982. In 1985, it concentrated on the research prospects that were opened up with the source. On the basis of the Swiss National Fund and additional expert advice, the Swiss Science Council recommended the realization of SINQ with a high priority. Therefore, the ETH Board included the project in its 1986 ETH funding proposal for buildings. SIN itself was to contribute significant monetary resources towards the construction of the source, but a new neutron hall for source plus instruments, as well as a neutron guidance hall for experiments with slow neutrons were included in the budget proposal. The Council of States approved SINQ in the fall of 1986 and the National Council followed in the spring of 1987.

After early difficulties, the SINQ was built in subsequent years under the project leadership of Erich Steiner. Because of the safety concerns of the regulatory body HSK, the plan of a liquid target was abandoned for the time being, and a conventional, water-cooled zirconium target was substituted. It was only in 2006 that PSI, in connection with a feasibility study of the accelerator-operated transmutation of reactor waste (MEGAPIE), had a lead-bismuth-eutectic target successfully built into the SINQ, which had as its result a doubling of neutron flux.

Just like the founding of SIN as a meson factory, the decision in favor of the SINQ had international effects. Just as before, the German Federal Ministry of Research and Technology (BMFT) decided to abandon the planned spallation neutron source in Jülich – a project in which Hans Willax, deceased in 1981, had participated – and to utilize the SINQ and the European neutron source in Grenoble for its researchers.

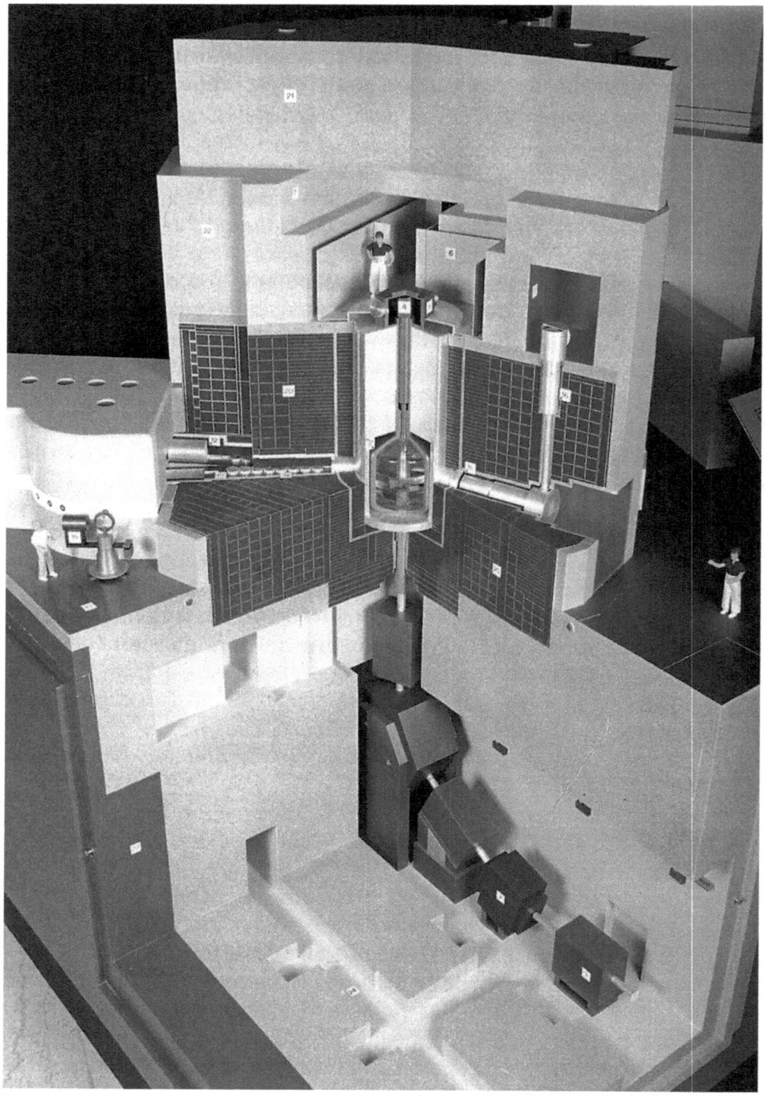

Fig. 60: Cutaway model of target block and beam guidance of the planned neutron source. Focusing and diversion magnets guide the proton beam through a vacuum tube and into the target from below. The target is housed in the center of a large tank containing heavy-water. Towards the left a beam tube runs to a neutron diffractometer with massive radioactivity shielding.

10. Actions of Science Policy outside of SIN

While the science policy initiatives described in the last chapter concerned the potential of the SIN facilities, SIN became active to an increasing degree outside its own area. In most cases, these initiatives were successful, albeit often for other participants, and only after the science policy establishment fought against most ideas until an understanding of serving scientific progress prevailed.

10.1 Accelerator Mass Spectrometry at the ETH Zurich

ETH President Ursprung decided to shut down the Tandem-van de Graaff accelerator in 1985. He reasoned that research in nuclear and particle physics could be conducted under the best of conditions at SIN and at CERN, and that this offered a welcome opportunity for cost reductions at ETH. However, the operators of the Tandem accelerator under Professor Willy Wölfli had already developed an attractive new application for the machine, i.e., Accelerator Mass Spectrometry (AMS), and had developed it into an internationally recognized top facility. This method proved to be important for archaeological dating and also for experiments concerning environmental research. Supported by EAWAG Director, Professor Werner Stumm, and by the well-known climatology researcher Professor Hans Oeschger of Bern, SIN management petitioned the ETH Board to continue operating the facility for the AMS research and to discontinue just the nuclear physics program. A compromise was found where SIN covered the expenses of the Tandem accelerator. Its nuclear physics program was discontinued and the machine was entirely dedicated to AMS. Their results were so successful that a successive facility was built in the 1990s.

10.2 RCA Laboratory

In 1986, General Electric acquired the RCA company. RCA had operated two research laboratories, one of them in the USA and one in Zurich. GE decided to concentrate its corporate research in their laboratory in Schenectady (USA) and to abandon RCA's two laboratories.

The RCA laboratory in Zurich was active in basic and in applied research in microelectronics and enjoyed a first-class reputation. It cooperated with ETH and oversaw many doctoral theses. Their research themes generally concerned the areas of semiconductor technology and information technology which were acknowledged as important and support-worthy. To simply shut down such a successful laboratory in Switzerland seemed absurd to SIN. Therefore, the SIN directors made a number of contacts, but discovered that no one at the appropriate ETH institutes, in industry, or at the appropriate institute in Neuchâtel was prepared to do anything. Since at the same time, Swiss industry and PTT, together with Federal offices, had begun an initiative for the creation of a national laboratory for compound semiconductors, SIN management suggested that the RCA laboratory be saved and integrated into the ETH domain.

SIN management made efforts, within the aforementioned initiative, to convince the ETH Board and all appropriate federal offices and succeeded in the fall of 1987 in getting the Confederation to acquire the RCA laboratory and to temporarily incorporate it into SIN. The question of its final structure and its incorporation was discussed with the ETHs and was to be decided by Swiss Parliament in 1988.

Thus, the laboratory in Zurich continued operations. The then-laboratory director, Karl Knop, contributed in large measure in that he succeeded to hold together his staff of proven scientists during the times of uncertainty. Knop also recognized the new demands on advanced semiconductor technology and was able to meet them in subsequent years. According to Parliament's 1988 decision, the RCA laboratory came under the aegis of PSI and later, when PSI needed funding for its large installations, of CSEM in Neuchâtel. A portion of the RCA laboratory, however, was transferred to PSI. The affected research groups conducted research projects in the area of nanotechnology in newly constructed clean rooms in Würenlingen which later became the basis for applications of SLS in nanotechnology.

Contract Uncertainty at the last Minute

In the fall of 1987, the decision of the Federal Council regarding the acquisition of the RCA laboratory had been prepared, but the contract for the purchase of the laboratory for $1 had not yet been signed. In charge at the ETH Board was the scientific advisor for annex institutes. He had discovered a clause in the contract which stated that the price for the laboratory was "one dollar and further consideration". What did "further consideration" mean? The official in charge called a law office in Zurich that was experienced in international economic questions and asked for advice. The attorney stated that if the government wanted to pay only one dollar, the clause should be changed to "for one dollar and no further consideration". The official in charge asked the GE attorneys for a change. When he entered a commercial building in the Frick Valley with Professor Blaser, who had been invited to explain SIN to the Aargau Chamber of Commerce, the official in charge was called to the telephone by the doorman. The call was from the GE attorney in charge of the negotiations in New York, explaining that the wording had to be this way in order to avoid unnecessary taxes in the USA. Would the ETH Board agree to a cover letter stating that GE would not expect more than one dollar for the laboratory? The matter had already been submitted to the Federal Council, there was no time to waste on further queries. The official in charge agreed and the contract was signed. At least Blaser was able to present the "purchase price" to the GE representative in the form of a valuable antique silver dollar coin, encased in a beautiful wooden box.

10.3 *Heating Reactor GEYSER*

In 1985, several concepts for heating reactors were created in Switzerland. The first one came from EIR. It was for a conventional, small light-water reactor. Industry, led by ABB, countered with a concept for a gas-cooled high-temperature reactor. Finally, the multi-talented SIN physicist Georg Vecsey, together with engineer Pal G.K. Doroszlai of Elektrowatt AG, suggested a technically innovative heating reactor – GEYSER – which was based on the geyser principle and functioned without active components, making it inherently very safe.

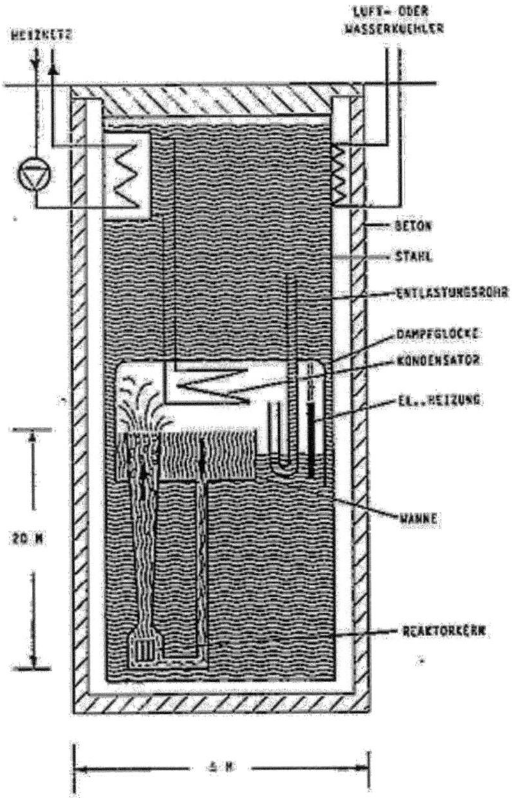

Fig. 61: GEYSER concept: The reactor creates hot steam in a steam generator, and then the steam is condensed again by a cooling system. The cooling system transfers heat into the district heating network.

This unusual undertaking was supported by the National Energy Research Foundation (NEFF) as well as by several private sector companies (Balduin-Weisser among them). However, it also presented a trigger for a conflict in science policy because it was argued that such a project did not belong to SIN's area of responsibility. Nevertheless, SIN management approved the construction of a test facility, which was supported with significant monetary resources from the other partners. The point was that this project would open new avenues which, technically speaking, had every reason to succeed.

In an evaluation of the three concepts, the GEYSER concept was judged to be the most promising. And, by the way, it became the subject of a dissertation at the ETH. None of the three concepts was pursued further. Although a user company by the name of Geysir AG had been founded for the GEYSER heating reactor, the technology transfer was not successful – probably also because widespread criticism of such a facility among the population made its construction impossible.

11. Future Large-Scale Facilities for SIN

Already in the year 1981, SIN asked itself questions about the future of the institute. They planned a new large-scale facility, again in the area of particle physic with possibly broad applications. This facility was intended to become a new focus of development in the 1990s.

The project was conceptualized in accordance with research needs in subsequent years. In the beginning a K meson (kaons) factory was being discussed. A concept named ASTOR was developed by Werner Joho in collaboration with Urs Schryber. It was presented in 1981. The "Accelerator and Storage Ring" could deliver protons between 2.5 and 4 GeV, depending on the magnets used, either in a continuous beam or in single pulses. The machine would be "fed" by the ring cyclotron but would use the 600 MeV proton beam only for short periods so that the spallation neutron source operation would not be significantly compromised. The ASTOR facility could become reality within SIN.

In subsequent years the concept was optimized several times. It could become an independent facility, or it could be utilized as a preliminary stage for a 20 GeV synchrotron for a European kaon factory (the so-called European Hadron Facility EHF). These concepts were developed by Joho and Schryber in 1984, based on the K factory at TRIUMF as a "High-Intensity Proton Synchotron (HIPS)". The facility consisted of several acceleration rings in a single tunnel of a 100-meter radius. In discussions regarding scientific utilization, even higher energies, such as 30 GeV, were targeted. Physicists at Swiss and European universities studied the research program in detail and judged it attractive. However, because of its high cost, such a facility could probably only be realized with international cooperation. Such cooperation was not on the horizon. Therefore, the project was abandoned in 1986 in favor of a B meson factory because it was thought to be useful to more external users.

Such a concept had been conceived in the mid-1980s by Ralph Eichler, together with Tatsuya Nakada and Klaus Schubert. B mesons are approximately six times heavier than protons and are the heaviest elementary particles to date on which still open questions of weak interaction can be researched. The B factory would consist of one electron-synchrotron and two separate storage rings for electrons and positrons of approximately 6 GeV. This would create optimal conditions under which to experiment in the areas of heavy quarks and heavy leptons. In addition, the facility could parasitically produce short-wave synchrotron beams for other fields of research. Contrary to ASTOR, the B factory would utilize primarily well-known components, making realization low-risk.

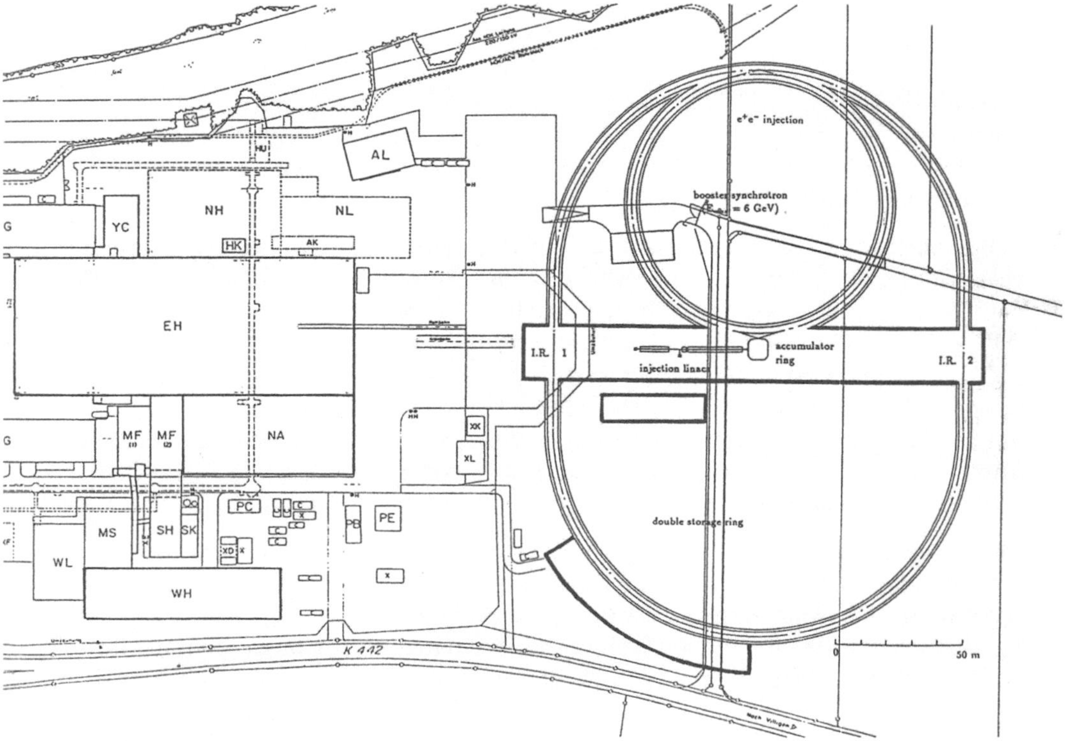

Fig. 62: B Factory. Layout of storage rings for electrons and positrons and their buildings on the SIN site.

In 1987, SIN decided to press ahead with the realization of the B factory proposed by Eichler. A work group made up of Swiss and German physicists produced a design which was to be submitted to the ETH Board for evaluation in 1988. The investment for the acceleration complex and for a universal particle detector ran to about 170 million francs. It suggested a bi-national development together with Germany. During a discussion with the German Federal Research Minister Heinz Riesenhuber in February 1989, Federal Councilor Flavio Cotti raised the subject of German participation. Riesenhuber made it clear that Germany would not participate in the investment costs. At most, it would make payments for future use of the facility.

In March of 1989, the ETH Board decided to consult regarding the project with the Swiss Science Council, the National Foundation as well as the universities. On the basis of these consultations, the Board decided during its session in September 1989 that the research range on this machine was too narrow and denied the application. Another decisive factor for the denial were the five national research programs that were the focus for the 1990s in the areas of environmental sciences, biotechnology, optics/optoelectronics, solid-state and materials research, which would require the commitment of significant financial resources.

After this denial, an idea was floated to cooperate with CERN and build a B factory in CERN's ISR tunnel. In a meeting between Blaser and the then-CERN Director Carlo Rubbia, they discussed the division of labor between CERN and PSI. In the words of Rubbia, this would be a so-called "Yalta Agreement". When Blaser asked "who is Stalin?", Rubbia replied that he was Charles de Gaulle (who had not been present in Yalta!). The idea was quickly abandoned again because a financial participation of PSI (in 1988, SIN had become part of PSI, see chapter 13) in CERN would certainly not have facilitated PSI's expansion.

As mentioned before, a synchrotron beam would have been created parasitically on the storage ring of the B factory which could have been used for experiments. Rafael Abela was working on this part of the project. This created the preconditions for SIN to immediately concentrate on the construction of a synchrotron radiation source based on Abela's work, and to appoint a design team consisting of Abela, Joho, Rivkin and Schryber. Rafael Abela played an important part regarding future experiments. The same went for Leonid Rivkin, who had been enticed away from the Stanford Linear Accelerator (SLAC) and named project leader by SIN for the machinery portion of the B factory.

Wilfred Hirt became active on the scientific and political sides in order to drum up support from Swiss universities and private industry, knowing that he could count on Nobel Prize winners Heinrich Rohrer and Alex Müller as driving forces. This was very important as the following experience illustrates: A group consisting of Wilfred Hirt, Rafael Abela, Bruno Reihl and Werner Joho traveled to Basel to solicit support from a chemical company for the synchrotron beam source – ideally financial participation, but at least a letter to the Federal Council recommending the realization of the project. Company management reaction was lukewarm. They explained that their employees were using synchrotron sources in the USA, and all it did cost was a few flight tickets to participate in this type of research.

A breakthrough for this project with the ETH Board and federal authorities finally came for the future PSI Director Meinrad Eberle. And only after international recognition came for PSI's synchrotron light source SLS as a top facility, the chemical industry in Basel wanted to be part of it.

Independent of the type of the new large-scale research facility, it became evident that its construction, in addition to the not yet completed high-current project and the construction of SINQ, would exceed the limits of SIN resources. This circumstance was not the primary reason for the merger of SIN and EIR into PSI, but the total budget of an institute of more than twice the size significantly expanded the possibilities of renewal.

12. The Collaboration with EIR

12.1 Infrastructure

The choice of location for SIN had had various reasons (see paragraph 2.4). One of them was the proximity of EIR, which already existed as an ETH annex institute on the opposite side of the Aare River.

As the initial construction project of SIN, the bridge across the Aare connecting both institutes was started in October 1969. It quickly gained regional importance. At first, the drinking water pipeline, which connected the reservoir of Würenlingen in Böttstein with the community, was constructed. In the 1980s, the same was done for the feeder and return flow lines of the district heating system for the lower Aare valley, REFUNA. In both cases the entire region profited by not having to invest in their own constructions. Most importantly, the regional population wanted to use the bridge for their traffic. SIN initially opposed general use of the bridge for liability reasons because radioactive materials were being transported between the institutes. A compromise was reached by allowing local traffic between the communities of Villigen and Würenlingen. Generally, the connection was marked as a private road.

Fig. 63: Construction of the Aare Bridge 1969, view of the east shore.

Already during the construction and installation phases of SIN, the infrastructure departments of EIR provided valuable technical and administrative assistance. They took on several projects, such as the provisioning of the heating and cooling water facilities as well as extensive planning work for the electrical facilities.

The two institutes concluded an extensive agreement in 1973 where all items of logistical cooperation were listed in numerous addenda. They covered cooling water supply, substitute medium-voltage power supply, workshop services, employee cafeterias and safety. It turned out early on that two cultures were colliding: The industrial and economic orientation of the EIR, based on its history, and SIN's orientation as a university institute. Generally, the cooperation functioned well whereby the EIR provided most of the services to SIN because it was the older institute with an existing infrastructure.

The REFUNA project became another infrastructure collaboration between EIR and SIN. It concerned the decoupling of thermal discharge from the nuclear power plant Beznau and its use as a district heating system for the lower Aare valley. At the founding of the REFUNA AG in 1983, the Swiss government became a shareholder through EIR and SIN contributing 250,000 francs each from their budgets for the purchase of shares and agreeing at the same time to buy their heating requirements from REFUNA. Equally important was the personal commitment of both institutes in technical and organizational questions. Primary players were Edmund Loepfe on the EIR side and Peter Schwaller and Christoph Tschalär on the SIN side. Loepfe took on the presidency of the REFUNA planning consortium and Schwaller, who was also mayor of Endingen, would later become president of the REFUNA AG.

Finally, when it became necessary to expand both employee cafeterias because of staff increases, the president of the ETH Board decided in the mid-1980s that both cafeterias were to be merged into one under a new common concept on the Würenlingen side of the river. That decision caused huge waves among the personnel of both institutes – at EIR because they had to forego table service, and at SIN because their personnel had to hoof it across the bridge for lunch. Nevertheless, the decision was realized under the project leadership of the scientific advisor for the annex institutes, and under PSI the personnel cafeteria developed into a most successful and generally accepted institution.

12.2 Research Projects

Soon cooperation between EIR and SIN loomed ahead also in the area of science. The isotope production marked the beginning. After commissioning of the first injector cyclotron, construction was begun in 1972 on an irradiation station for the SIN isotope production. The EIR produced isotopes primarily for medical applications in their reactors. SIN, with its acceleration facilities, was able to produce low-neutron isotopes which were in much demand in nuclear medicine and biology because of their low patient radiation exposure and excellent possibility of detection. They expanded EIR's product range and could be integrated into the existing outbound logistics. SIN produced such isotopes for the first time in 1974. In 1975, routine production of iodine-123 was begun. Additional isotopes of rubidium, krypton, fluorine and xenon were also produced. The cooperation was expanded in subsequent years. The demand for isotopes caused SIN to plan for isotope production with Injector II in 1978. Production began in 1987. Altogether, significant investments were made by SIN in technical facilities, and by EIR in an isotope building with special laboratories for the preparation.

In 1982, the cooperation was deepened in the area of spallation neutron source (see chapter 9). SIN was able to draw on EIR specialists in the areas of process engineering and thermodynamics. The most important partners for this project, though, were the scientists of the Laboratory for Neutron Scattering at ETH, which was located on the EIR site and used its reactors.

During the same year, there was collaboration on the PIREX project (see chapter 9). EIR also began to logistically support Swiss groups at CERN, just like SIN already did.

12.3 *Common Visions for the Future and Hayek Studies*

By the mid-1980s both institutes had drawn close in the following project areas:

- Material Studies with neutrons and protons
- Nuclear medicine
- Nuclear fusion technology
- Collaboration with high-energy experiments at CERN

In spite of the fact that some of these areas lay outside the two annex institutes' main focus of work, they were relevant because they led to new scientific applications. In this situation, both management groups decided to strengthen their own cooperation. They planned periodic joint management sessions, mutual representation in advisory committees and the investigation of a closer cooperation through a work group. All of this was designed as a means of building trust between two management groups from different cultures. In concrete terms, no effective results were achieved by these measures.

On the initiative of Wilfred Hirt at SIN, the possibility of a merger of EIR and SIN had been in early discussion. These discussions assumed that SIN would have to expand its almost 20-year-old facilities soon with a new, large-scale project in order to retain its good international reputation. Whether the institute would be able to raise the necessary funding, was uncertain. At the same time it was felt that EIR , in view of the fact that Swiss industry was not contemplating any larger developments in nuclear technology, could soon be viewed as too large by federal authorities. For both institutes this meant that, as separate entities, they might face real dangers of cutbacks, or to even be shut down as superfluous. That this danger was real was confirmed with several, temporarily unofficial, remarks by politicians.

In 1983, both management groups had drafted a joint report to the attention of the ETH Board president, which described possible future developments. Beginning in 1984, the Board also began asking questions about a joint future for EIR and SIN. The ETH Board named a committee for this purpose which showed no results for the time being. For the most part this was due to the fact that SIN saw an important opportunity – even a necessity – in a merger, while on the EIR side a conservative, even defensive attitude prevailed. Nevertheless, there were important superiors at EIR actively involved in opinion making; above all main department manager Edmund Loepfe, who, by the way, suggested naming the merged entity "Paul Scherrer Institute" when the difficult question of what name to choose was being batted back and forth.

In 1985, the ETH Board asked Nicolas Hayek's company Hayek Engineering to conduct an optimization study covering the entire ETH domain. According to the master plan, Hayek conducted two subsequent studies for EIR and SIN. The first study was completed in 1985. It examined streamlining possibilities in infrastructure. It recommended that both institutes take on certain tasks where they were especially efficient, such as EIR for technical services and SIN for materials logistics. Hayek estimated personnel savings at about 30 positions. In March 1986, ETH Board directed both management groups to carry out the streamlining measures recommended by Hayek and to funnel excess positions into ETH reserves. Although both management groups began merging infrastructure parts, the conclusion emerged that, because of the differences in group cultures, an effective streamlining of infrastructure would be possible only within a merged institute.

The second study was conducted in 1986 and was concentrated on new orientations for both institutes. For this purpose, the ETH Board retained four high-ranking experts, all of them physicists, i.e., Pierre Aigrain, State Secretary of Research in the French government; Wolf Häfele, Board President of the Nuclear Research Facility at Jülich, Germany; Geoff Manning, Director of the Rutherford Appleton Laboratory in England; Jean Teillac, French High Commissioner of Atomic Energy. As his representative, the ETH Board President deputized his scientific advisor for the annex institutes, Andreas Pritzker. Together with Hayek Engineering, these four experts examined the research programs of both institutes and pondered their future direction and optimal structure. They submitted a comprehensive expert report in April of 1986. In their report of May 1986, which Nicolas Hayek sent with a cover letter to ETH Board President Maurice Cosandey on June 4, 1986, Hayek Engineering AG summarized the experts' conclusions as follows:

"The recommendations are aimed at an intensive collaboration and a merger of EIR and SIN as best possibilities for the future. According to the experts' own experiences, research laboratories with several major strategic missions have the best prospects of a long-term, secure and useful scientific-technical work platform which is marked by constant inner renewal, dynamic and attitude of innovation. With a single strategic mission, institutes have a hard time justifying their reasons for being long-term. .... A merger of EIR and SIN, with possible resizing and certainly restructuring of both institutes, appears to be the optimal solution for securing their future and for preserving the worthwhile qualities of EIR and SIN, especially since both institutes have already begun joint or complementary research projects that can be expanded. The merger and reorientation require first-rate central management in order to complete them quickly and successfully."

ETH Board President Professor Maurice Cosandey, who was planning to retire in February 1987, together with his designated successor, Professor Heinrich Ursprung, came to the conclusion that with this merger a good opportunity should be embraced. The ETH Board concurred. Its resolution was leaked to the press. In an article dated July 19, 1986 the Tages-Anzeiger already announced the merger under the headline "Marriage of convenience in Swiss nuclear research?" A subtitle stated: "Even after EIR and SIN merger, no exit from nuclear energy." And the article began as follows:

„Are the Eidgenössische Institut für Reaktorforschung (EIR) in Würenlingen and the Schweizerische Institut für Nuklearforschung (SIN) in Villigen, both of them so-called annex institutes of the Eidgenössischen Technischen Hochschulen (ETH), being merged? The ETH Board wants to unite both partners in a marriage of convenience and better integrate the institutes into the university realm."

The article was remarkably well researched. It mentioned that EIR Director Heini Gränicher was not at all happy with the decision of the ETH Board. He would have preferred to expand the EIR as a national energy research center and was hoping that the Federal Council would not follow the expert advice of the Hayek study. SIN Director Jean-Pierre Blaser, on the other hand, fully supported the ETH Board decision, and SIN had a great deal less merger anxiety than EIR. When asked if the decision was meant to set the course for a direction in energy policy, ETH Board General Secretary Fulda replied that the ETH Board had oriented itself strictly on the basis of science. Even a merged institute would not be able to completely dispense with reactor technology. The directors of both institutes shared that opinion.

The ETH Board informed the Federal Council of its intention to merge the institutes. The Council confirmed receipt on September 10, 1986 and asked the ETH Board to submit suggestions for the mandate, organizational structure and operations of the new annex institute within six months. In order not to miss the tight deadline, President Cosandey named a project organization on September 29, 1986. It consisted of a steering committee and a project team. Michael Kohn (president of the EIR advisory committee) headed the steering committee whose members were Urs Hochstrasser (director of the Federal Office of Education and Science), Eduard Kiener (director of the Federal Office of Energy), Verena Meyer (professor of physics at the University at Zurich), Heinrich Ursprung (president of the ETH Zurich), Bernard Vittoz (president of the EPF Lausanne). In an advisory capacity to the steering committee were the directors of EIR and SIN, the general secretary of the ETH Board, Johannes Fulda, as well as Andreas Pritzker in his capacity of scientific advisor to the annex institutes. The project team was headed by Jean-Pierre Blaser. The team consisted of Heini Gränicher (EIR director), Wilfred Hirt (SIN's deputy director), Edmund Loepfe (manager of the scientific-technical department of EIR), and again as advisor, Andreas Pritzker.

The project organization worked purposefully on its mandate and consulted with outstanding scientists of both institutes. The steering committee was able to submit its report to the ETH Board in March of 1987. In April 1987 the ETH Board petitioned the Federal Council to merge EIR and SIN into PSI, effective January 1, 1988. The Federal Council ordered a broad legislative process by consultation to be completed by September 30. Positive comments prevailed. According to the general secretary of the ETH Board, Johannes Fulda, this paved the way to the "largest merger within the Federation since the founding of the Swiss Federal Railroads (SBB)".

Fig. 64: Both institutes SIN and EIR

Accordingly, prior to the adoption of the resolution by the Federal Council, it came to an extensive process of communication sharing among the various Federal Departments since many Federal Agencies had an interest in the formation of the new institute. The Federal Council passed the merger resolution on November 30, 1987, but asked the ETH Board to submit a report on the work program priorities and the medium-term resource planning for PSI for approval by the end of the year. Politically, this created a new course because the Federal Council demanded balance between nuclear and non-nuclear energy research – a subject that would keep PSI management busy for quite some time.

13. The Beginning of the Paul Scherrer Institute (PSI)

After Jean-Pierre Blaser, as project team leader, had prepared for the merger of both institutes, the Federal Council elected him logically to serve as the new institute's first director. He served in that capacity for almost one and a half years, at which time he became eligible for retirement. Deputy Director was again Wilfred Hirt.

How did the PSI get its name?

The ETH Board, the project organization and the federal authorities completed the founding of PSI on an ambitious timeline. Sometimes this required some fire-fighting. The project organization had suggested "Paul Scherrer Institute" as the name. The ETH Board agreed, but there were opinions that the use of a person's name was unusual in Switzerland, and that the Federal Council should demand a neutral name. However, there was no opposition on the part of Federal Agencies during the preparation of the resolution. Prior to the Federal Council's passing of the merger resolution, the scientific advisor to the annex institutes had to contact the only surviving daughter of Paul Scherrer's at the last moment and ask her permission. The daughter was happy to grant her permission on the phone and promised to post a letter to that effect immediately. The letter reached the files for the Federal Council session just in time.

The PSI consisted of the research departments of Nuclear and Particle Physics (headed by Prof. Roland Engfer), Biosciences (Edmund Loepfe), Solid-State/Materials Sciences (Prof. Hans-Rudolf Ott), Nuclear Energy (Hans-Peter Alder), General Energy (Prof. Peter Suter), and as additional infrastructure the Technical-Scientific Department (Christoph Tschalär) and Administration (Andreas Pritzker). Another member of the management team was Professor Meinrad Eberle as coordinator for energy research. Additional management team members were Karl Knop as head of the former RCA laboratory, Ralph Eichler as head of the Project B Meson Factory, and Walter Fischer as head of the Project Spallation Neutron Source. Assistants to the directors were Karl Buob, Peter Schwaller and Martin Jermann.

The following tasks were tackled during the initial years of the PSI:

- Merging SIN and EIR programs and projects into the newly-founded research fields of Bio- and Materials Sciences.
- Merging and re-sizing the infrastructure.

- Re-sizing of the original primary fields of particle physics and nuclear energy in order to free resources for new development proposals.
- Development of new research fields, such as non-nuclear energy.
- Integration of the RCA laboratory into the PSI.
- Continuation of the large-scale projects SINQ and high-current expansion.
- Evaluation of a new large-scale facility after the rejection of the B Meson Factory.
- Development of new management processes.

The ETH Board expected a streamlining effect from the merger of the administration and infrastructure departments of both institutes. This effect had been estimated at about 30 positions by the first Hayek study; however, the budget of the new institute was cut only moderately since its new directions required additional monetary resources. This affected especially the development of non-nuclear energy which was now politically expected.

A press conference held in March 1989 regarding the institute's direction found much echo in the media. As expected, interest was focused especially on the balance of energy research and less on the at least as important expansion of materials sciences, which would be successfully completed by PSI in subsequent years with its methods of structural analysis of materials.

With any merger of organizational entities, there will always be winners and losers, and the losers have the tendency to derail the organization – if only just to show that they were right to be skeptical. There were two communities within PSI which – partially justified – feared the waning of their influence. The users of SIN feared the loss of their focus, reduced resources, as well as loss of the university culture of SIN. They could be calmed down somewhat due to a lot of convincing done by Jean-Pierre Blaser and Wilfred Hirt. Circles interested in EIR – electricity industry, machine industry, administration and politics – assumed that the PSI would move closer to the universities and feared a loss of influence over the programs. Warned by internal losers, the safety authorities feared a deterioration of safety standards under the former SIN directors – totally unjustified since the SIN's facilities had made the highest demands on safety.

The result of all these concerns was a demand on PSI management to retain the traditional structure for each stakeholder and to reserve funds. Especially problematic were the demands made by governmental agencies because they could put pressure on the PSI via their Federal Councilors and the ETH Board. As is usually the case, all these stakeholders had their representatives or allies within the institute.

This turned the initial years into a balancing act for management. Everywhere workable compromises had to be found without losing sight of the overall future direction. Looking back, it had been a wise move to name an experienced director as head of the merged institute, someone who was familiar with both previous institutes. The choice of an outsider would probably not have led to success. The failure of Jean-Pierre Blaser's successor, Anton Menth, who resigned after a short time, is proof. A stroke of luck was the eventual selection of Meinrad Eberle as successor to Menth. With his strong personal commitment he succeeded to calm down the skeptics and to win the trust of political, economic and scientific factions, so that the PSI was able to develop most successfully.

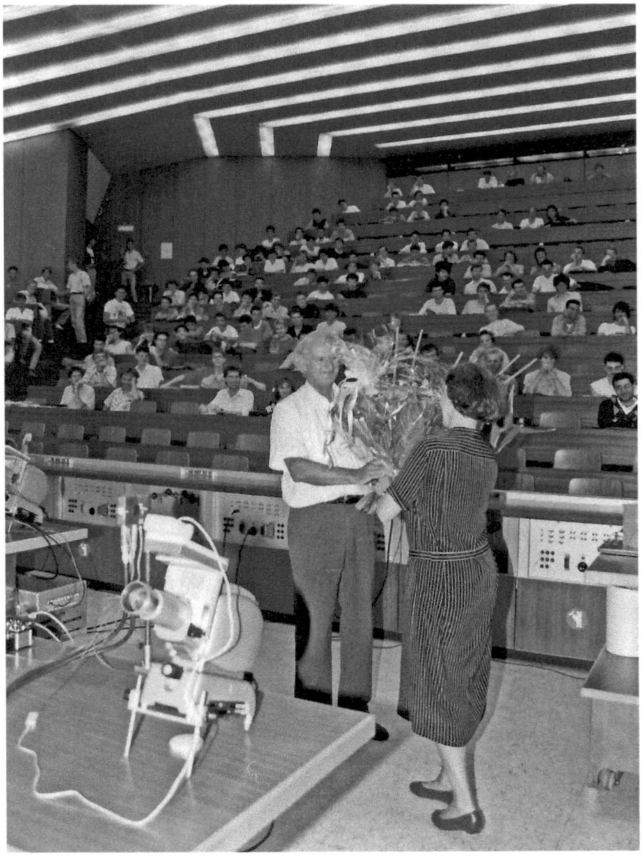

Fig. 65: Farewell to Jean-Pierre Blaser after his last lecture at the ETH in July 1989, conducted by Christa Markovits, his long-term lecture assistant.

The retirement of Jean-Pierre Blaser from the Human Resources Department's view

When Jean-Pierre Blaser retired as director of the PSI, there was hardly any farewell ceremony. There was one at the ETH whose Physics Department had been strongly influenced by Blaser over many years. Perhaps the reason was that Blaser maintained a small office at PSI for some time because he still had certain scientific functions, such as the representation of PSI at the "Heureka" Scientific Exposition. This kept him visible although he did not interfere in PSI's business affairs. His retirement was announced to the PSI staff the same way as all retirements: In the newsletter "PSI-aktuell" appeared the following entry without further details in the list of retirees: "Blaser, Jean-Pierre" plus the effective date of retirement of March 31, 1990, and below that the sentence: "Our heartfelt thanks to the named colleagues for their valued work and our best wishes for their retirement."

14. The Significance of SIN to Science and Science Policy

The basic idea behind the founding of SIN in the 1960s was to concentrate country-wide efforts on the expensive research in nuclear and particle physics without interfering in the federalist structure of Swiss research.

In the roughly twenty years of its existence, SIN was able to point to top performances in research as well as the technological development of its facilities. For the research, which was primarily carried out by external users, SIN's facilities offered the possibilities of new precision measurements in the areas of nuclear and particle physics and to examine occurrences that were forbidden by the standard model. Various constants of nature could, therefore, be delimited more precisely than had been possible before. Structural analysis of solid-state materials and liquids with muons also showed successes.

On his retirement, facility users presented Jean-Pierre Blaser a substantial collection of four volumes containing several hundred publications of work that had been conducted at SIN facilities in the 1970s and 1980s.

Between 1974 and 1987, about 240 projects were published in nuclear physics, about 50 in particle physics, about 60 in atomic and molecular physics, as well as about 140 in muon physics. During the heydays of medium-energy and muon physics, about 50 articles were published per year, consistently in well-known international journals, and often the authors specifically thanked the SIN. Remarkable are also the collaborations that can be seen in the list of authors.

Quite a few of these publications were probably also connected with dissertations. For the education of doctoral students SIN proved itself very attractive. It offered an alternative to CERN whose experiments were often attended by hundreds of authors and where individual scientists, especially doctoral students, got lost in the crowd.

Also the applications – cancer therapy with particles, positron emission tomography, super-conductivity, the spallation neutron source – can all be classified as very successful. And finally, the Synchrotron Light Source Switzerland (SLS), which was built by the successor institute PSI, had its beginning through a SIN initiative.

Important and more technologically scientific contributions of SIN were the development of cyclotrons and detectors. An estimated ten dissertations were written about machine development. The builders and operators of SIN's – and later PSI's – accelerators were successful in reaching world-class output when compared with other facilities.

Just the realization of SIN's accelerator facilities, as originally conceived, could not be taken for granted. While it is true that experiences with other similar facilities could be utilized, SIN's facilities represented a pioneering achievement. For a long time, it seemed possible that they could not be realized at all, or realized as conceived. That the specialists at SIN were successful in boosting capacity to a high degree is due to their professional competence, paired with self-confidence. Thanks are owed to management for attracting highly qualified specialists, and for providing an environment in which their ideas could become reality.

Fig. 66 : In the control room 1974: Jean-Pierre Blaser, Hans Willax (at the logbook) and Urs Schryber (at the control device) while fine-tuning the proton beam.

As summarized by Werner Joho, it was due to four measures – improved beam quality from the injector, flat-top cavity, excentrical injection, and increased accelerator voltage thanks to copper cavities – that the accelerator physicists were able to boost the beam current from an originally planned 100 microamperes to a momentary value of 2,400 microamperes. In spite of this massive increase, the absolute losses on extraction fell by a factor of 10 to 0.5 microamperes. That allowed the produced radioactivity to be kept within limits. It also permitted, to this day, easy manual maintenance of critical cyclotron components without having to use complex manipulators. There are many accelerators with higher energy than the 590 MeV of the ring cyclotron. However, its medium beam capacity of 1.4 megawatt is an astonishing world record to this day, especially when one considers that the SIN (now PSI) machine has been in operation since 1974!

Equally unusual was the extent to which SIN management became engaged in science policy during its 20-year existence. Unusual because committees and authorities responsible for science policy view the management of an institute – after a possible hearing – more as political objects than teammates. As the years went by, SIN successfully launched a number of initiates that would have effects way beyond the institute and were often useful to other institutes (see chapter 10).

A hardly measurable success of the SIN concept lay in the fact that the institute brought together scientists from Switzerland and from many European and non-European countries to collaboratively conduct experiments. They conducted them each in their respective scientific specialties. However, since SIN was working on a broad spectrum of physics and beyond, it was able to gather together a number of different scientific disciplines. And the development and operation of their large-scale research facilities demanded close cooperation among scientists, engineers and technicians.

That such a complex operation as SIN, where always new technical and organizational problems demanded solutions, was able to function efficiently and productively, could also not be taken for granted. Decisions often were based on difficult technical and scientific factors, so that standard management techniques were of little use. As Director Jean-Pierre Blaser expressed it in his 1984 annual report (the 10th anniversary of the beam operations): "Central to this operation are people with a high degree of professional knowledge and a willingness to excel. That these traits are so highly developed at all levels of our institute, is due to each individual employee. For that I wish to express to all my admiration and my thanks."

A Personal Retrospective by Jean-Pierre Blaser

I The Elements of Success of SIN

Self-organization and lean management

During the development of the SIN facilities, special problems arose due to its size and its new acceleration principle of much higher beam capacity which differed from anything that had been done before. All parts of the project – particle dynamics, magnets, high-frequency system, vacuum system, shielding and radioactivity safety, and of course the research plans as primary objective – were so tightly interrelated that it was necessary for all participants to have sufficient theoretical and technical knowledge of the entire project. The customary division of labor in large-scale industrial projects among specialized engineers working on successive prototypes did not apply here because the so-called prototypes had to be fully functional right away.

This led to an organizational structure that created itself quickly around a number of specialists who, with their broad theoretical and technical knowledge and strong leadership skills, automatically took on managerial functions. These were mostly young physicists with no experience but great enthusiasm and competence who tackled all the unique questions that arose. It created a natural hierarchy which – unlike in other organizations – was able to dispense with seniority, diplomas, titles or the status of experts.

All of this allowed SIN to develop an extremely lean management with very little "red tape". Special thanks go to Wilfred Hirt for his organizational talent. He established a method where all decisions were previously prepared by those responsible in informal discussions until all agreed on goals, processes and possible problems – he referred to it as creating "unité de doctrine". Since all thought alike, the realization of decisions proceeded amicably, and the usually necessary protocols and instructions could be dispensed with.

Unusual Employees

This work environment made possible a number of unusual careers for employees who initiated important projects at SIN with their innovative ideas. Here are just a few of numerous examples:

1. Walter Fischer started out in the Theory Group. Later he became shift leader of the accelerator operations, and finally the initiator and implementer of the spallation neutron source SINQ. Thanks to his competence and personal engagement, paired with considerable perseverance in organizational and political matters, he was able to achieve significant scientific and technological successes.

2. Charles Perret belonged to the group that dealt with concepts and construction of beam guidance systems, target stations and experimental facilities. There, he developed into a leading expert in the difficult problems of radioactivity, shielding and radioprotection. During his work he recognized the possibility of using the proton beam from Injector I for the treatment of melanoma of the eye. He succeeded in organizing the medical cooperation and to build the technically innovative OPTIS machine, the first truly successful therapy project with particle beams in Europe.

3. Georg Vecsey developed quickly and independently into the leading specialist for cryogenics and super-conductivity at SIN. He understood the enormous potential of this innovative magnet technology and developed especially original concepts for the muon channel and the Piotron irradiation machine. With his large-scale SULTAN facility he set international benchmarks for the technological development of super-conductive materials. And on the side, he invented, together with an engineer partner, the innovative GEYSER heating reactor.

Fig. 67: Jean-Pierre Blaser with mechanic in the shop

Such blossoming of personal competence and motivation at SIN was not limited to scientific-technical fields. Ms Christine Bertsch, for instance, independently developed her secretarial work into a professional travel bureau. And in addition, she catered to the needs of many guest scientists who came to SIN from all over the world, with great human sensitivity and devotion. That was not always easy. For instance, how does one organize travel, accommodation and living conditions for a Japanese physician sent to SIN for the study of particle therapy, who perhaps finds himself with his family in a German speaking Swiss village? There were scientists from the Soviet Union to take care of, who had to find their way around our free culture, but also doctoral candidates with financial problems. The grateful guests would often declare: "The most important person at SIN for me was Mrs. Bertsch!"

Supportive, not inhibiting government authorities

Especially lucky for SIN was the understanding and support from supervising authorities, first from ETH President Hans Pallmann and Vice President Claude Seippel, later from Heinrich Ursprung in his function as ETH president, then as president of the ETH Board. He, too, was no friend of bureaucratic methods, as the following example illustrates:

After Ursprung had decided to shut down the Tandem-van de Graaff accelerator in Höngg for budgetary reasons, but later agreed to SIN's suggestion of continuing the operation under partial financing by SIN in order to conduct accelerator mass spectrometry, he declared that it was enough to hang a sign at the entrance to the building, saying "AMS Laboratory of SIN". With such disarming decisions, not many administrative regulations would be necessary.

Understanding Federal Agencies

That such matters could be solved quickly and pragmatically was also due to the great goodwill SIN enjoyed for many years by the Federal Finance Department.

Our human resources management was complex and in large parts did not conform to the categories of the Federal personnel regulations. There were assistant positions, grant subsidies for doctoral candidates from other universities – often from poorer countries, scholarships for stays abroad, short or longer invitations to guest scientists from various countries. Also, SIN often needed to quickly retain expensive specialists for the construction of the facilities. Since much development was conducted under a specific project, which had a beginning and an end, it was clear and undisputed that many positions had to be temporary.

Once the facility was operational, the large investments made it necessary to run the accelerator as well as the experiments 24 hours a day, 7 days a week. The only exception was the shut-down during Christmas which made it possible to let the radioactivity subside before the big annual shut-down. This caused demands for shift work and on-call repair services. SIN also had to make available accommodations in the guest house, meals, as well as access to the shop, to material storage and to transportation around the clock. Federal regulations for capturing hours worked with a time clock, overtime and time compensations did not turn out to be the most logical solutions. That such an operation still functioned was due to the fact that everyone participated with enthusiasm and enormous commitment. Hardly anyone considered the status of government official desirable, even though it was possible. The Finance Department showed understanding for the special circumstances, also because the SIN employees agreed, and allowed the institute to proceed.

Fig. 68: The new control room, November 1987. The control room was manned around the clock.

The support of the Finance Department was also important when it came to procurement. Sometimes orders worth multiple millions had to be placed, and when specific technologies were needed, they had to sometimes be placed with foreign companies. Because of the demanding specifications, which sometimes needed to be adjusted to developmental changes, often difficult situations would arise with disagreements about guarantees and the fulfillment of specs. One such typical example was an order to a specialized company in California for a series of beam guidance magnets. The magnets were ready, but before they could be shipped, the company declared Chapter 11 bankruptcy, and delivery was blocked by the American justice system. For a research institute to solve that from small, far-away Switzerland without any in-house attorneys was not exactly trivial, but it worked. And here, too, there were no accusations from the Finance Department, just support, especially from their attorney Dr. Schneeberger who held SIN in high regard.

Safety not as an end in itself

When it came to matters of safety, SIN also aimed for simple procedures. There were many hazards considering the extremely heavy components, such as magnets and shielding, strong magnetic fields, high-frequency generators of high power, vacuum systems and, of course, especially the radioactivity generated by intensive particle beams inside the accelerators, beam guidance systems and target stations. Many potential hazards were closely interrelated and differed depending on each operational situation. Such complex situations did not make it easy to guarantee safety with the conventional methods of extensive regulations with threshold values, instructions and prohibitions. Therefore, management depended more on the technical knowledge and experience in research operations of their staff in order to recognize and judge the real hazardous situations. This also eliminated the well-known effect where over-regulated provisions often forbid an obviously nonhazardous behavior, which tends to blunt safety consciousness. The supervising authority – the Federal Office of Health – was also satisfied with SIN's procedures. This office was also in charge of CERN where a radiation safety policy was pursued that was good but also as liberal as possible.

Summary

Taking an overall view, SIN's successes were due to the engagement and commitment of its staff as well as the goodwill of external entities that all preferred pioneering achievements and processes, within reasonable limits, to the type of over-regulation that chokes off all initiative.

II The Appropriate Positions of Science Policy

The successful development of SIN demonstrates clearly how important it is to make the proper science policy decisions.

The ETH Board wants to maintain the high standards of the ETH

It began with the appointment procedure for Scherrer's successor. ETH Board President Pallmann desired to maintain the international reputation of the ETH which was due to Scherrer's initiative in nuclear physics and accelerator technology. The cyclotron planning team represented the basis. However, in the appointment conversations I had with Pallmann, he did not explain to me what I had to do to nurture the inheritance from Scherrer. With his remark "the ETH wants horses that can pull" he signaled that large-scale projects would also be welcome. Pallmann just knew what a good university needs for research: Innovation, not necessarily conservation. With that he gave me a free hand to proceed with particle physics.

The step towards particle physics is obvious

The first new direction, in which I was able to participate, was to turn away from pure nuclear physics, which was being researched intensively everywhere else, because the physics institutes of larger universities were able to operate the necessary facilities themselves. While it was clear that particle physics at top energies could only be conducted in large-scale research centers of important countries, such as the USA and the Soviet Union, or at international laboratories such as CERN, the energy range up to 1,000 MeV, later called medium-energy physics, appeared to be especially interesting. The first machine at CERN, the cyclotron, had already yielded important results in this area so that working with mesons offered interesting applications to other scientific fields.

This situation had led to a suggestion by the ECFA (European Committee for Future Accelerators), to build a meson factory, an accelerator facility whose high-capacity beams of pions and muons could be used as tools. Two such projects were already in development in the USA and in Canada, and the Federal Republic of Germany also had begun initial investigations.

Switzerland dares to start a meson factory

The second step in the process was the suggestion to build such a facility in Switzerland under the aegis of the ETH. Some people considered this a sign of arrogance, if not to say delusions of grandeur. But preconditions were favorable and CERN's location in Geneva had a positive influence.

The next step in the process was taken because of the size of such an undertaking. Either the Federation would have to create a large-scale, specialized research institute, or the undertaking could take the direction of a national research center which would be built and operated by the ETH, but would scientifically "belong" to all Swiss universities. Because of Switzerland's federal university structure, this idea caused quite a bit of opposition, but also enthusiasm, especially among the smaller universities and the western part of the country.

This meant that the institute to be developed was clearly conceived as a user laboratory where the research would be conducted by external users. But not exclusively. Claude Seippel, President of SIN's Building Committee, supported management and decided that SIN itself would also need to be given a research mandate. This resulted in mixed research groups with amicable relations. The local SIN staff was especially important to these groups when it came to the development of measurement equipment and the necessary beam facilities. Nevertheless, there were many research projects that were conducted by university institutes.

After much deliberation, Villigen was chosen as the location. This decision was quite correct because it emphasized the national character and also made the future merger with the EIR into the PSI possible.

Of course, the most important decision was the choice of accelerator. Since the requested concepts from industry were not convincing, and imitating the concepts of Los Alamos (Linac) or Vancouver (H-minus ions) seemed hardly suitable for financial and spatial reasons and also for capacity reasons, the concept devised by Hans Willax gained traction. It represented a totally new cyclotron concept, compact and therefore economical, with large capacity potential, albeit risky because it was a new development. Long-term, the decision in its favor would have a very positive impact. By the year 2014, the accelerator will have brought great benefits to research for forty years.

A large-scale undertaking creates opponents and requires decisive action

How did the periphery react to the decisions? The strong opposition by some universities was mentioned above. But there also was opposition within ETH. The nuclear physicists were afraid that the change of research venue to a larger center would endanger the historically successful research at their own institute. And the solid-state physicists were almost startled at the tought of such an undertaking. Private industry, which had always been very interested in ETH activities, acknowledged the project as attractive, but thought that the research goals were not in alignment with economic interests.

Obviously, this did not make it easy for the ETH Board president to present the project as desirable to the ETH Board and the responsible Federal Council. What would be decisive in this regard was the opinion of the theoretical physicists who judged the dynamic development away from nuclear and towards particle physics as very important from the standpoint of science policy. They stated their position clearly, and especially the commitment of ETH Professor Res Jost carried a lot of weight.

Finally, it was Federal Councilor Tschudi who, with full confidence in ETH Board President Pallmand and in the project leadership, took personal responsibility to present the large-scale project to the Federal Council and then to Parliament and to ensure that financing would be available. Viewed today, it was remarkable that Tschudi, a social democrat, supported with personal conviction the construction of a Swiss Institute for Nuclear Research, just like social-democratic Federal Councilor Spühler obtained approval for the construction of nuclear power plants under great personal engagement, in order to prevent the burning of fossil fuels for the production of needed energy.

Science policy is a rocky road, but one should never give up

After this unique and for our country unusually large project found favor in Parliament and the budget proposal was passed, the construction of SIN proceeded swiftly. All participants dealt with the significant technical demands of developing the accelerator facility and the construction of new beam guidance systems and research equipment in an admirable manner with their great expertise and initiative.

The most important task of institute management was to influence science policy with their determination of new research goals and research techniques. From the very beginning, their central idea was that, in addition to the current particle physics research programs of a meson factory, applications in other sciences should be initiated early. As described above, these were in particular the application of muons in solid-state physics and chemistry, neutron scattering with a spallation source, radiotherapy with pions and protons, as well as the use of PET in nuclear medicine. International conferences connected to these initiatives were organized by SIN early and created appropriate international contacts. The initiatives were accepted with great interest, they even led to science policy decisions in foreign countries, but in Switzerland, the scientific "establishment" often could be induced to participate only after successes could be seen.

Sometimes it was the young researchers who had a braking effect. For instance, nuclear and particle physicists who used SIN were of the opinion that the new injector was not needed because the available beam quality was enough for them.

And that the coordination among institutes is often lacking, could be seen in the example of CERN. It was true that the ECFA had recommended the construction of a European meson factory as successor to the much weaker CERN synchrocyclotron. However, after successful commissioning of the SIN facility, the "Scientific Policy Committee" at CERN – I was a member at that time – raised the question of a possible shutdown of the synchrocyclotron. Suddenly, there were all sorts of reasons why they could not dispense with the machine. Malicious gossipers cited arguments such as "the library is better at CERN than at SIN", and "SIN's travel expense reimbursements are lower"...

As these examples show, science policy is a rocky road, but one should never give up. Often, it's not enough to just do what the "establishment" wishes. But even democratic voting among the researchers themselves would often lead to an incorrect result.

Fig. 69: ... a rocky road: Construction site 1970

The researchers are important for opinion making

The users played an important part when it came to opinion making, especially their various committees which were actively participating in the experiments when it came to allotment of beam time. In addition, international committees, such as the Scientific Committee at SIN, were important advisors. This created many happy professional and personal contacts, including those numerous "after-hour sessions" to which Frauke Blaser would always issue such liberal invitations. Nevertheless, these valuable committees always stayed in an advisory capacity. Decisions and accountability remained with institute management.

The use of SIN facilities ran beautifully; the number of proposed experiments always exceeded available beam time and international interest in the use of SIN was especially high. This very much impressed ETH Board President Jakob Burckhardt, who had diplomatic experience, and congratulating me, he said that SIN had truly become a "réussite" (success).

Research subjects also must end

At the beginning of the 1980s it became clear that the primary research subject, which had opened SIN, was coming to an end after 20 years and that a new and even larger project would be needed if the institute wanted to remain tops in the game internationally. This was true because the expenses were way too high to be conducting just local research.

As previously mentioned, the idea of a kaon factory was seen as an attractive possibility by many users. Kaons are indeed of great interest in particle physics, but their short half life make beams, like they are possible for applications with pions and muons, hardly possible.

Then Ralph Eichler suggested a B meson factory based on an electron accelerator. The particle physics questions were, of course, of great importance. The downside was the relatively narrow research area, so that already in early discussions the term "synchrotron radiation" could be seen on the blackboard. That's why the group assigned to study the electron accelerator also investigated its use as a "light source".

The decision as to which of these two projects to present to the ETH Board was a difficult one. Users favored the kaon factory while institute management judged the B meson factory as the scientifically more important one. Since the latter also fit well into the additional use as a synchrotron radiation source offering usefulness in wide-ranging applications, the decision was made in its favor. The kaon factory project, pursued by TRIUMF in Vancouver for many more years, ended up not being approved.

SIN applied for the B meson factory, but the ETH Board did not approve it. The reasons given were the relatively narrow research area as well as the lacking interest on the part of private industry. Therefore, the research into important questions of particle physics could no longer be the primary task at SIN.

A user laboratory for new research areas

As originally conceived, SIN's goal remained the operation of a large-scale research facility for Swiss and international users with ties to universities. Envisioned were applications for as many different scientific specialties as possible. Added to the already existing facilities or those in development for muons, neutrons, accelerator mass spectrometry, and several applications in medicine, the important possibility of research with synchrotron radiation followed naturally. The group that had been tasked with the development of the electron accelerator of the B factory could be used directly for the large-scale project of a "Swiss Light Source". Getting agreement under science policy would turn out to be more difficult. Some universities rejected the large-scale facility because of expense. Thanks go, above all, to Wilfred Hirt that – based on his understanding of scientific principles and his great political skill – he was able to obtain the urgently needed approval of the National Science Foundation.

As far as large-scale research facilities are concerned I was of the opinion – not always shared by all – that in view of the merger with EIR, the engineering type of research at such an institute should also concentrate on large-scale facilities. In energy research, for instance, this would be research facilities for nuclear reactors, large solar mirrors or hot laboratories connected with a research program for nuclear waste management. For smaller experiments, where the financial resources of university institutes are sufficient, university groups are more innovative, less expensive and more efficient for academic training.

Summary

Historically speaking, science policy has always been very inconsistent, and I think it will remain that way. Let's remember the suppression of Galileo or the story of Leonhard Euler, one of the greatest mathematicians of all time who, in spite of being supported by the also distinguished Bernoullis in his home town of Basel, was rejected for a position at the university after many applications. Euler's incredible life's work was made possible only through the science policies of then-rulers. Euler ended up first at the Academy of Peter the Great, and then he was invited to Berlin by Frederic the Great, and finally called back to Petersburg by Catherine the Great. Who is being supported today by those with power and money?

For our contemporary times, SIN was an impressive example of science policy: How in a small country with a short and un-bureaucratic decision making hierarchy, the right people were always in the right place, always willing to decide and always granting scientific freedom. Well, it was a piece of good fortune and reason to be grateful.

Appendix 1: The Facts about SIN

Total Investments

As stated in paragraph 2.4, the initial investments in the SIN facility until 1973, including site acquisition, totaled 137.5 million francs. These were credits applied for in the ETH building budget proposals of 1965 and 1972 as well as a contribution by the National Science Foundation for Injector I. In subsequent years, the Swiss Parliament approved additional investments in SIN under ETH building budget proposals of: 15.3 million francs for Injector II (1978); 7.4 million for the central building (1981); 11.3 million for the extension of the experimental hall (1983); 32.6 million for the spallation neutron source (1987).

Construction projects costing less than 2 million francs were not applied for through building funding proposals, but were financed out of the annual budget of the Office of Federal Buildings . Between 1980 and 1984, this amounted to (in million francs):

- Side wings of the assembly hall (1.0)
- Expansion of the SULTAN Hall and Helium Compressor Station (0.6)
- Shop and storage building (1.9)
- Medical and SULTAN building and expansion of medical building (2.1)
- Isotope laboratory wing (1.8)
- Expansion of supply channels (0.9)
- Various smaller remodeling projects (2.8)

Added to this is the building donated by the Cancer League for the Piotron of approximately 2 million francs.

Beginning in 1974 there were additional credits for machinery, approximately a total of 163 million until 1987. These were used for the expansion of the accelerator facilities, but also for the constantly changing needs of the experimenters in experimental areas as well as the corresponding research facilities, such as secondary beams and detectors.

In summary, the total investment for the buildings and the facilities of SIN between 1966 and 1987 ran to roughly 380 million francs.

The following Fig. 70 shows the SIN site development.

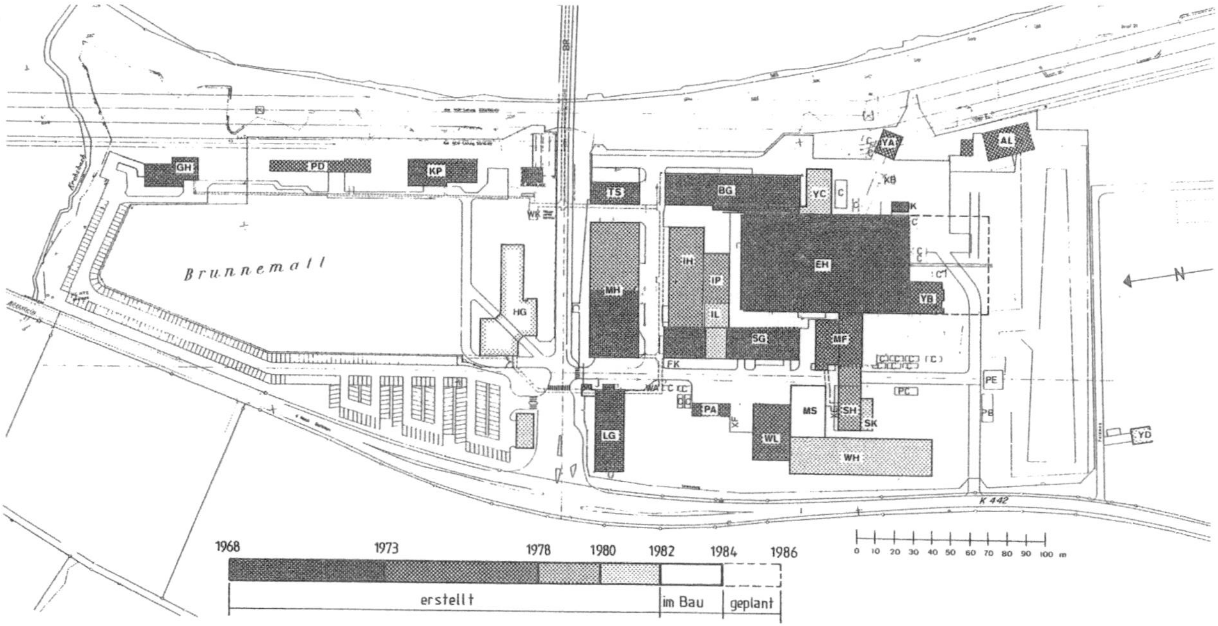

Legende der Abkürzungen für die Gebäude

AL Aktivlager
BG Betriebsgebäude
EH Experimentierhalle
GH Gästehaus
HG Zentralgebäude
HK Helium-Kompressor
IH Injektor-II-Halle
IL Isotopenlabor
IP Isotopenproduktion
KP Kantinen-Pavillon
LG Laborgebäude
MF Medizinisch-biologisches Forschungsgebäude
MH Montagehalle
MS Medizin- und Sultan-gebäude
PA Bauleitung, Hausdienst, Transportdienst, Post
PB Werkstatt Hausdienst
PD Labors und Büros
PE Werkstatt Hallendienst und Hausdienst
SG Speisegebäude
SH Sultan-Halle
SK Sultan Helium-Kompressor
TS Transformatorenstation
WL Werkstatt- und Lager-Gebäude
WH Werkstatt- und Lager-Halle
YA nE1-Areal
YB pM1-Areal
YC Niederenergie-Areal C
YD Experiment Neutrino-Masse (Uni Zürich)

Finances

Until 1973, the resources for the construction of SIN were handled as a project credit by the Office of Federal Buildings.

It was only in 1974 that SIN was given its own operating budget. The operating expenses shown in the table below (in round numbers) describe the following situations unique to SIN as compared with other ETH institutions and other research centers:

- The relatively high proportion of investments in equipment is characteristic for an institute whose mandate is the construction and operation of large-scale research facilities. The jump in 1980 in the overall budget as well as equipment credits is due to an additional credit of 9.3 million francs with which the Federation financed the iron blocks lent to SIN for shielding in order to offset a shortage in the Swiss war material reserve.
- Also characteristic is the personnel cost proportion of 50% or less of total expenses. In other ETH Institutes this runs closer to 75%.
- Typical for SIN was the number of employees with limited contracts. This is due to the fact that a different professional profile is needed for the construction of the facilities, as opposed to their operation. This allowed SIN management to remain flexible when it came to personnel expenses, unlike institutes where new employees are hired on a permanent basis or even as government officials. Financing limited-in-time employees out of the general budget item "Teaching and Research" was normal for ETH institutes. Since SIN saw itself as a university institute they claimed the same right. For a long time this was viewed with suspicion by unions who even demanded government official status for the staff.
- The rise in total expenses is largely due to an increase in head count.
- In the beginning, revenues consisted of contributions from the Federal Republic of Germany (considerable) and Austria (more symbolic) as payment for the use of the SIN facilities. Later on, contributions followed for research projects, for instance, for LCT and SULTAN. Revenues for medical therapies were a problem because they flowed directly into Federal coffers because they were not considered research projects. SIN solved the problem by creating a foundation. At first this only applied to the building that had been donated by the Cancer League to house the Piotron. Later on this was expanded so that SIN could take the therapy fees and then spend them on research projects.

Jahr	Gesamtbudget [MCHF]	Maschinen [MCHF]	Personal [MCHF]	Mitarbeiter gesamt	Mitarbeiter nichtständig	Einnahmen [MCHF]
1970	-	-	-	160	n.a.	-
1971	-	-	-	201	n.a.	-
1972	-	-	-	224	n.a.	-
1973	-	-	-	261	61	-
1974	32.1	12.9	10.5	274	60	2.6
1975	30.8	10.3	11.1	283	62	2.8
1976	33.2	10.1	11.5	289	63	3.9
1977	33.6	9.1	11.7	298	71	4.9
1978	35.2	10.0	12.1	306	77	4.8
1979	37.4	11.4	12.7	320	91	5.6
1980	48.1	21.1	13.5	333	104	5.6
1981	39.8	11.4	14.5	342	109	5.5
1982	42.0	11.4	16.0	357	n.a.	5.3
1983	46.0	10.7	16.6	369	140	5.1
1984	48.4	11.5	17.3	375	144	5.4
1985	47.5	11.4	17.9	369	138	5.2
1986	48.0	11.1	18.6	377	144	5.2
1987	50.9	11.5	18.5	427	196	5.2

Fig. 71: Annual expenses of SIN 1970 to 1987. Prior to 1973, the expenses were covered by the construction budget and personnel costs were covered by ETH.

[Translation: Gesamtbudget = overall budget; Maschinen = facility investments; Personal = personnel costs; Mitarbeiter = number of staff, gesamt = total, nichtständig = limited contracts; Einnahmen = revenues]

Apprenticeship Training

SIN valued apprenticeship training in various technical professions. In 1981 they had 8 apprentices, and already 13 in 1983. By 1987 that number had risen to 16.

Publicity Activities

SIN was the first ETH domain institute that started a systematic PR program. Based on the premise that the institute had been financed by Swiss tax payers and, therefore, belonged to them, a program of public tours was organized. All types of groups with an interest in the institute received a tour of the facilities with an explanation of the equipment and the research that was being conducted with it. Scientists, technical employees as well as members of SIN management acted as tour guides. These tours contributed a lot to the appreciation the regional public showed SIN. They created contacts between scientists and lay people and forced the former to speak in laymen's terms about their work.

Facility Operation

Optimal usage of the significant investments in the SIN facilities – and soon the onslaught of experimenters – demanded 24-hour operations seven days a week. Part of this need was driven by the fact that the accelerator warm-up took several days. The operations personnel worked in three daily shifts. Wherever shifts could be dispensed with, on-call services were introduced which saved expenses.

The work of experimenters was not regulated because, for the most part, they commuted from somewhere else.

Each Christmas the facilities were shut down so that the generated radioactivity could subside. After the Christmas Holidays there was a shut-down of several weeks' duration for servicing, planned repairs and renovation.

Prior to commissioning Injector II, every fourth week was dedicated to low-energy operations. After Injector II, the ring accelerator could be operated continuously.

Appendix 2: Organization of SIN

Organizational Charts and Personnel

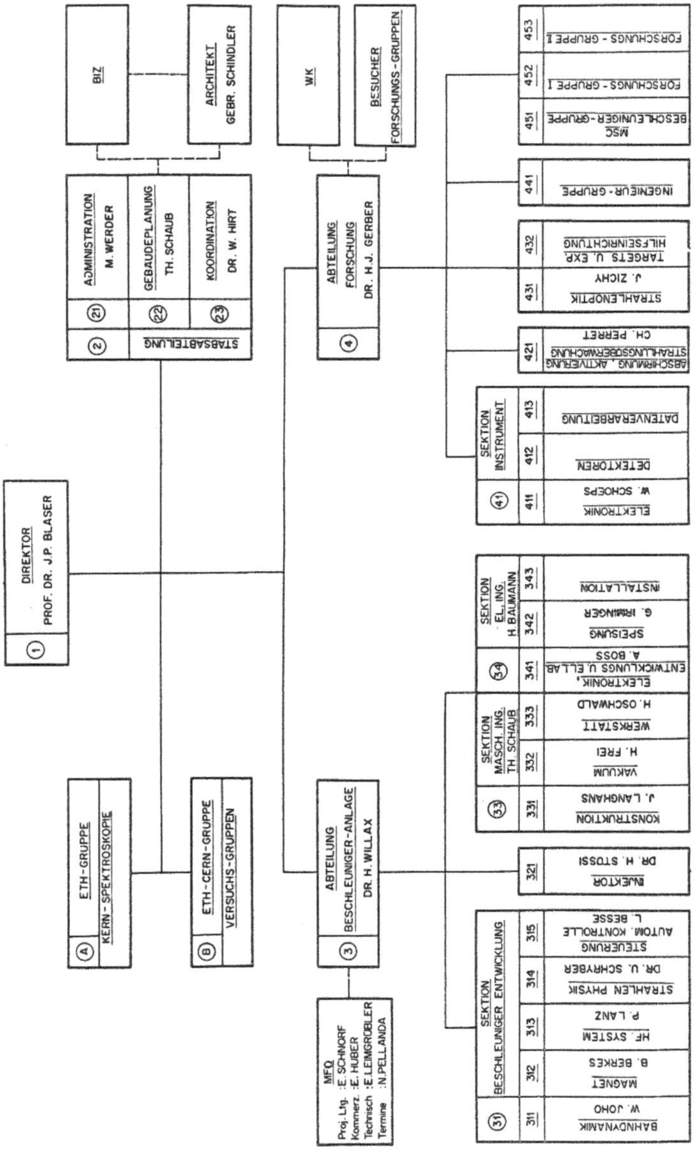

Fig. 72: Organizational chart 1969

In 1969, SIN had about 100 employees. Most of them were employed at ETH. There was a temporary organizational scheme that was appropriate for project work, see Fig. 72:

In 1970, SIN already had 148 employees, plus an additional 12 from the ETH and from universities as well as from private companies. At this time, the status of SIN personnel had not yet been determined. The intention was to integrate the personnel into the Federal administration. To make that possible required a suggestion for an organizational structure that would be valid long-term.

Therefore, a sub-committee of SIN's building committee, consisting of Professor Peter Preiswerk of CERN and Fritz Grütter of BBC (who previously worked at CERN and later changed to Motor Columbus Engineering), suggested a rough structure for the organizational flowchart according to Fig. 73:

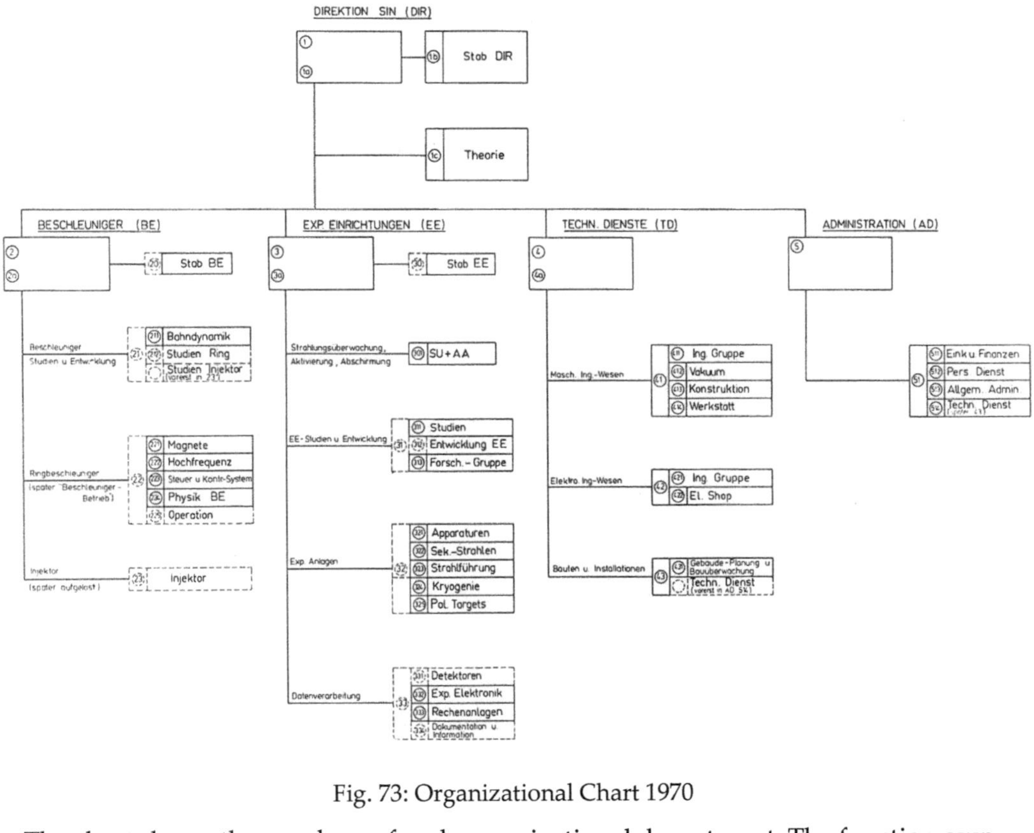

Fig. 73: Organizational Chart 1970

The chart shows the numbers of each organizational department. The function owners are shown in the next Fig. 74.

telle im rganigramm	F u n k t i o n	N a m e
1	Direktor SIN	Blaser J.-P. Prof. Dr.
1a	Stellvertreter	Willax Hans Dr.
1b	Direktionsassistent	Hirt Wilfred Dr.
1c	Leiter Theorie	Scheck Florian Dr.
2	Leiter BE	Willax Hans Dr.
2a	Stellvertreter	Schryber Urs Dr.
211	Bahndynamik	Joho Werner Dr.
221	Magnete	Berkes Branko
222	Hochfrequenz	Lanz Paul
223	Steuer- und Kontrollsystem	Besse Ludwig
224	Physik BE	Schryber Urs Dr.
3	Leiter EE	Gerber Hans-Jürg Prof. Dr.
3a	Stellvertreter	Tschalär Christoph Dr.
301	SU + AA	Perret Charles
313	Forschungsgruppe	Pepin Mark Dr.
321	Apparaturen	Tschalär Christoph Dr., a.i
322	Sekundärstrahlen	Frosch Reinhart Dr., a.i.
323	Strahlführung	Guignard Gilbert Dr.
324	Kryogenie	Vécsey Georg
325	Pol. Targets	Mango Salvatore Dr.
332	Exp. Elektronik	Schöps Wilfried
333	Rechenanlagen	Pepin Mark Dr., a.i.
4	Leiter TD	Willax Hans Dr., a.i.
4a 1	Stellvertreter Masch. Ing.-Wesen, Bauten und Installationen	Schaub Theo
2	Stellvertreter Elektro Ing.-Wesen	Baumann Hans
41	Chef. Masch. Ing.-Wesen	Schaub Theo
	Stellvertreter	Langhans Jürgen
412	Vakuum	Frei Hans
413	Konstruktion	Langhans Jürgen
414	Werkstatt	Oschwald Hans
42	Chef. Elektro Ing.-Wesen	Baumann Hans
	Stellvertreter	Irminger Gottfried
422	Elektro Shop	Boss Alfred
43	Chef. Bauten und Installationen	Schaub Theo, a.i.
	Stellvertreter	Mollet Emil
431	Gebäudeplanung und Bauüberwachung	Mollet Emil
5	Leiter AD	Werder Max
51	Leiter AD	Werder Max
	Stv. Personalbelange	Kijewski Cyrilla
	Stv. AD ausser Personalbelange	Nussbaumer Josef
511	Einkauf und Finanzen	Nussbaumer Josef
512	Personaldienst	Kijewski Cyrilla
513	Allgemeine Administration	Haller Felix
514	Technischer Dienst	von Ballmoos Willy

Fig. 74: Organizational Chart 1970

In 1972 the Technical Department was put under the leadership of Urs Schryber, while Assistant to Management Wilfred Hirt took over the leadership of Administration on a temporary basis. After the departure of his predecessor, Max Schumm was named personnel manager.

In 1976 Frieder Lenz took over the Theory Group from Florian Scheck.

In 1977 the Federal Council issued a new ordinance to SIN which would become effective in 1978. This ordinance asked SIN management to reorganize the institute effective April 1, 1978. The most important changes concerned:

- The new function of a deputy director as well as a staff position "Research". The former was taken on by Wilfred Hirt, and the latter by Hans-Jürg Gerber, who passed the management of the experimental facilities on to John Domingo.
- The creation of a new department "Applications/Projects" to be managed by Urs Schryber. The primary task for the time being was the development of Injector II. Management of the Technical Department moved from Urs Schryber to Christoph Tschalär.
- The creation of a new department "Management/Administration" which combined the appropriate services. The new manager was Peter Schwaller.
- The creation of the Medical Project under the direction of Professor Carl F. von Essen. His mandate was the development of cancer therapies with pions.

In addition, a personnel committee was elected whose first president was Walter E. Fischer.

1981 saw additional changes in the organization. After the death of Hans Willax, Urs Schryber took over the Accelerator Department which was now again responsible for the development of Injector II. Because of new and increased activities, two projects were created: "Super-Conductors" under Georg Vecsey as well as "Spallation Neutron Source" under Walter E. Fischer. Both projects were larger programs requiring an increase in personnel.

In 1985, there was a change on the management level. The director of the medical projects, Professor Carl von Essen, who had successfully led the initial phase of the project, left SIN. Richard Greiner became his successor. In addition, SIN took over the group "Accelerator Mass Spectrometry" (AMS) from the ETH. The group continued to be headed by Professor Willy Wölfli and remained at its ETH Höngg location (see chapter 10). Finally, the department "Experimental Facilities" was restructured into the SIN Research Department (SINFA), with the goal of increased activity by SIN scientists in large-scale experiments.

In the fall of 1986 Frieder Lenz, having been offered a professorship abroad, ceded management of the Theory Group to Milan Locher. In 1987, the former RCA Research Laboratory was merged into SIN. It continued to be managed by Karl Knop who became part of the SIN management team.

Many people involved in the development, construction and later operation of the SIN facilities have not been specifically named in the textj152

. To list them all would exceed what can be covered in this book. We will make do with a portion of the photograph taken September 1973 (page 66) and list the names which could be elicited later.

Fig. 75: A majority of PSI personnel is assembled around the ring accelerator in September 1973.

Who is in the photo?

(List of personnel from upper left to lower right – lists assembled by W. Joho, U. Schryber and D. Brombach)

Fourth Row:

L. Dorra, Ch. Tschalär, C. Petitjean, NN, F. Adamec, NN, A. Brunner (NZZ), N. Straumann (Uni ZH), H. J. Gerber, P. F. Meier, H. Oehninger, E. Mollet, E. Steiner, Evi Madlener, J. Zichy, G. Irminger, G. Hauswirth, R. K. Ray, Erika Huber, F. Reinath, Schwarz, U. Hofer, NN

Third Row:

R. Kramer, G. Agardi, Steiner (Uni ZH), A. Janett, P. Sigg, R. Häussler, H. Leber, W. Keller, P. Lanz, L. Rezzonico, I. Chini, T. Schaub, U. Schryber, J. Duvoisin, W. Rothacher, A. Grüebler, H. Fischer, R. Bidler

Second Row:

H. Meier, NN, E. Tremp, R. Balsiger, H. Baumann, E. Aebli, R. Reimann, N. Bezic, A. Paulin, NN, NN, B. Bischof, S. Adam, N. Schmid, L. Besse, Missaghi, R. Hasan, M. Märki, R. Kaegi, J. Chérix, H. Einenkel, P. Rudolf, P. Jokinen, F. Beck, A. Jucker, R. Illi, P. Meyer, J. Collins, Plüss, NN,R. Schumacher, NN

First Row:

U. Rohrer, E. Wagner, T. Blumer, Ch. Perret, C. Perotto, NN, NN, NN, E. Mariani, E. Marcolongo, NN, I. Jirousek, P. Ziegler, E. Baumann, H. Haller, A. Boss, NN, H. Fahrenholt, Hedi Naef, P. Gross, NN, Heidi Pozzato, Beatrice Burger

Behind the machine:

D. Brombach, W. van der Hoef, D. Babic, J. Waleffe

On top of the cavities:

M. Olivo, H. Willax, W. Joho, T. Stammbach, H. Zubler, U. Kalt, H. Marti (BBC), Christa Markovits, H. Frei, R. Brunner, Schiesser, A. Rölli

Scientific Committees

Scientific Committees were an important part of the SIN organization. They brought an outside view, thereby serving quality assurance.

SIN had a Scientific Committee begining in 1966. Its members are described in chapter 7. Its membership would change over time. New members were Ch. Haenny (EPUL) and D. Maeder (University of Geneva).

Following a new Federal ordinance, an additional Advisory Committee was formed in early 1978, just like for the other ETH annex institutes. It consisted of people from circles interested in SIN and reported to both the SIN director as well as the ETH Board. The following eight people were named as members by the ETH Board: Prof. Ernst Heer (President, University Geneva), Prof. Urs Hochstrasser (director of the Federal Office for Education and Research), Hans Lauri (Federal Finance Administration), Prof. Christian Schlatter (University Zurich), Prof. Norbert Straumann (University Zurich), Prof. Gerard Wanders (University Lausanne), Prof. Anselm Citron (University Karlsruhe), Prof. Florian Scheck (Mainz). As usual, there were changes in such committee membership as the years passed. Additional members were Professor Christian Trepp (ETH), Professor Beat Hahn (Bern), Professor Gisbert zu Putlitz (Heidelberg), Bernard Ecoffey (Finance Administration) as well as Anton Menth (BBC). The Advistory Committee took over for the Scientific Committee which continued in a smaller format and was to deal with questions regarding SIN's future. That committee now consisted of Professors Florian Scheck (Mainz), Hedi Fritz-Niggli (Zurich), Jean Kern (Fribourg), Lukas Schaller (Fribourg), Roland Engfer (Zurich), Hans Blattmann (SIN and University Zurich), Gerrit W. Barendsen (Rijswijk), Otto Kofoed-Hansen (Cophenhagen), Alfred Seeger (Stuttgart), Colin Wilkin (London). They represented various fields of research that were being conducted at SIN. Here, too, membership changed as years passed. Additional members were Professors Ingo Sick (Basel), Ch. Michel (Zurich), H. Müller (Zurich), J. Deutsch (Louvain), F. Foroughi (Neuchâtel), Willy Haeberli (Madison), Wolfgang Pohlit (Frankfurt/Main), Peter Truöl (Zurich) and W. Baltensperger (ETH).

In 1985, SIN simplified its committee structure with the introduction of just one user assembly with two sub-groups. The intent of this structure was a closer cooperation between the different fields of research as well as more say for the users regarding the future development of SIN.

The Transition to the Paul Scherrer Institute

In January 1988, SIN was merged with EIR to become PSI. The next Fig. 76 shows the organizational chart of PSI from its first annual report. It makes the merger visible. The fields of research F1 and F4A are in accordance with the original missions of the two institutes. The fields of research F2 and F3 group the applications that had been developed in both institutes over the years. The field of research 4B finally – it also goes back to past initiatives of EIR – represents the developmental area "Non-Nuclear Energy Research". In both infrastructure areas B8 and B9, the first one especially concerns the large-scale research facilities which PSI had taken over and would develop further.

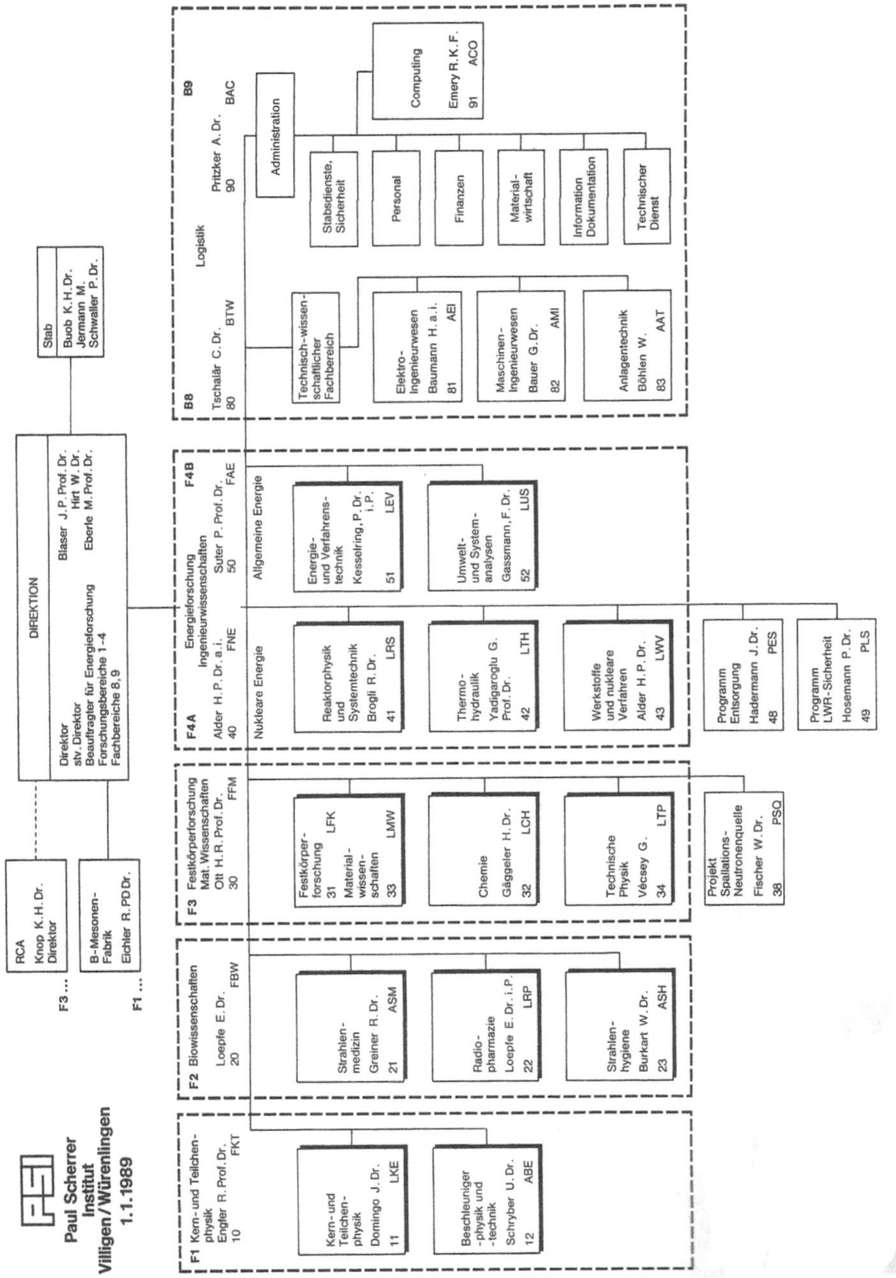

Fig. 76: Organizational Chart of PSI 1988

Appendix 3: The Development of the Experimental Hall

The development of areas within the experimental hall of SIN proved that SIN was adapting to the needs of the users and was willing to make significant investments.

The following areas M and E represent the (high-energy) targets for the pion and muon beams for which the areas were developed.

In 1969, only three low-energy areas A, B and C were planned while the high-energy targets were still in development. Soon areas B and C would be combined.

In 1970, SIN physicists planned the following high-energy areas: πM1 to πM4, πE1 to πE3, μE1 and μE2 as well as pM1 and two neutron areas. πM4 was eliminated in 1972. The following areas remained:

- πM1 Beam of highest resolution and high energy
- πM2 Beam of low energy
- πM3 Beam of medium resolution and high energy
- πE1 Beam of high resolution and highest energy
- πE2 Beam for stopped pions
- πE3 Beam for biology and medicine
- μE1
- μE2
- pM1 Beam of polarized protons
- nE1 Beam of non-polarized neutrons
- nE2 Beam of polarized neutrons

The areas πM2 and nE2 were not developed. Therefore, this list is in accordance with the hall layout in Fig. 21, where Area πE2 could also be used for muons.

The following spectrometers were planned or in development as early as 1973 because research operations were to begin in the summer of 1974:

- Pion Spectrometer
- Crystal Spectrometer
- Gamma Pair Spectrometer

The Fig. "1975" (this and the following figures are taken from SIN's annual reports) shows the hall layout at the end of 1975. It already contains planned areas outside the experimental hall. The iron frame along the western wall had farsightedly been installed in such a way that the proton beam could later be guided to a "west area". Fig. "1976" shows the experimental hall as of September 1976. Sketched in are the pion and muon areas with their respective spectrometers.

An isotope production facility at Injector I is operational as early as 1976. Also, area πE2 became area μE4 because of great demand for muons.

Fig. "1982" shows the experimental hall in 1982. The following facilities have been added in the meantime:

- Target πKS with crystal spectrometer for the so-called Gatchina experiment for the investigation of pion atoms.
- The bio-medical facility (BMA) with the Piotron.
- The new low-energy area C.

In addition, Injector II, including pre-accelerator, is in development.

Fig. "1985" shows the experimental hall as of 1985. New are the OPTIS equipment for the treatment of cancer of the eye, as well as an area for the PIREX experiment. Additionally, a second isotope production facility had been built at Injector II.

In Fig. "1987", showing progress by 1987 (prior to merger with EIR), the new large-scale detector SINDRUM is shown in area μE1. The experimental hall was being lengthened. In its southeast corner the ATEC area was installed in which radioactive components could be treated.

Finally, there was an experimental hall plan of the future in existence since 1984, Fig. "Zukunft". Sketched in are the neutron source SINQ with the neutron hall and neutron guidance hall, also the large LEPS spectrometer in area πE3.

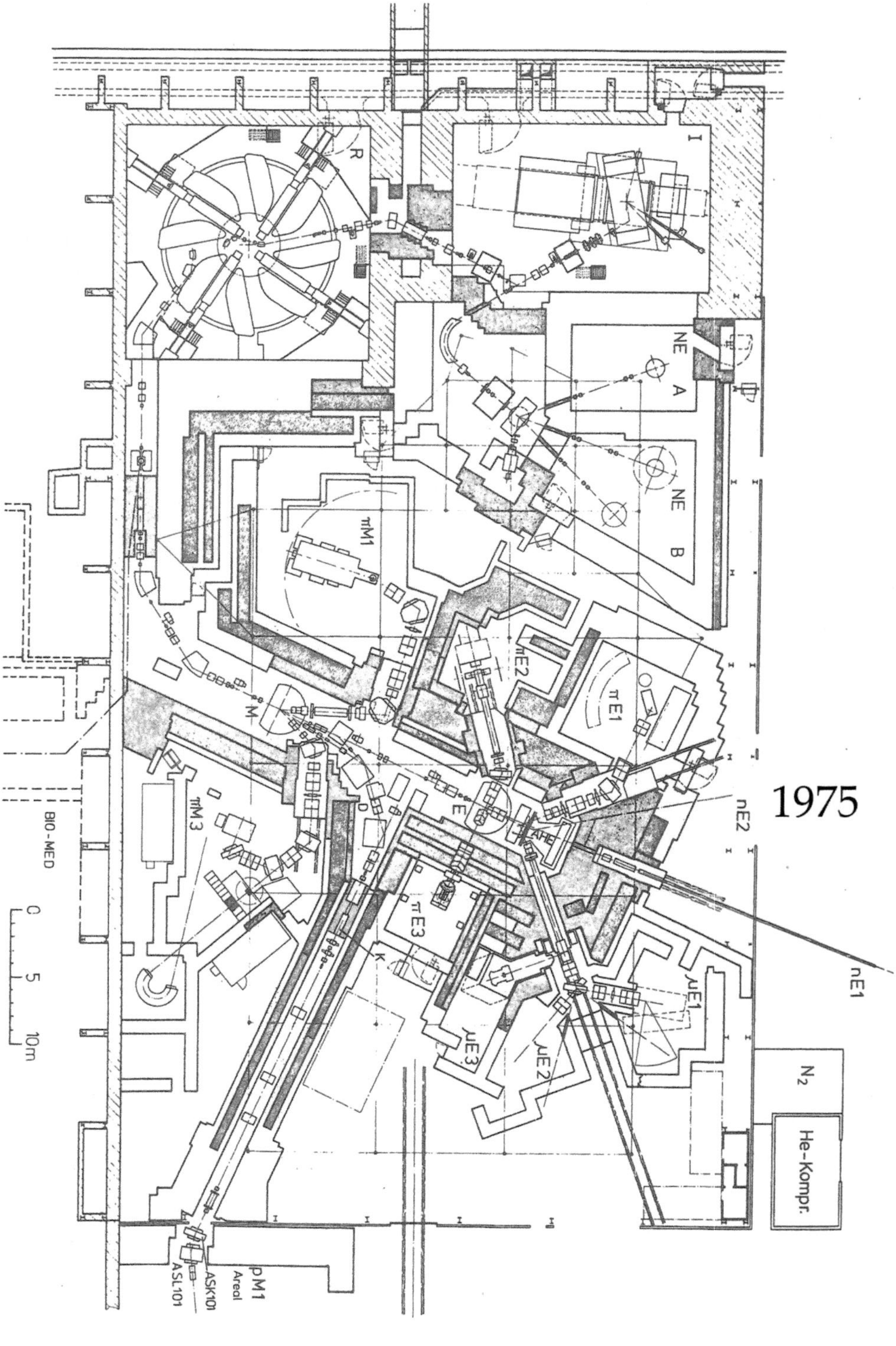
R
I
NE A
NE B
πM1
πE2
πE1
M
πM3
BIO-MED
E
D
K
πE3
μE1
μE2
μE3
nE2
nE1
N_2
He-Kompr.
pM1
Areal
ASK101
ASL101
0
5
10m

1975

SIN-Strahlführungssystem (September 1976)

1 Injektorzyklotron
2 Ringzyklotron
3 Analysierbunker
4 Niederenergieareale
5 Isotopenproduktion
6 Protonenkanal
7 supraleitende Müonenkanäle
8 Pionenspektrometer
9 Paarspektrometer
10 Kristallspektrometer
11 Target M (dünn)
12 Target E (dick)
13 Pionen
14 Müonen
15 Neutronen
16 Strahlstopper
17 polarisierte Protonen
18 biomedizinische Experimente mit Pionen

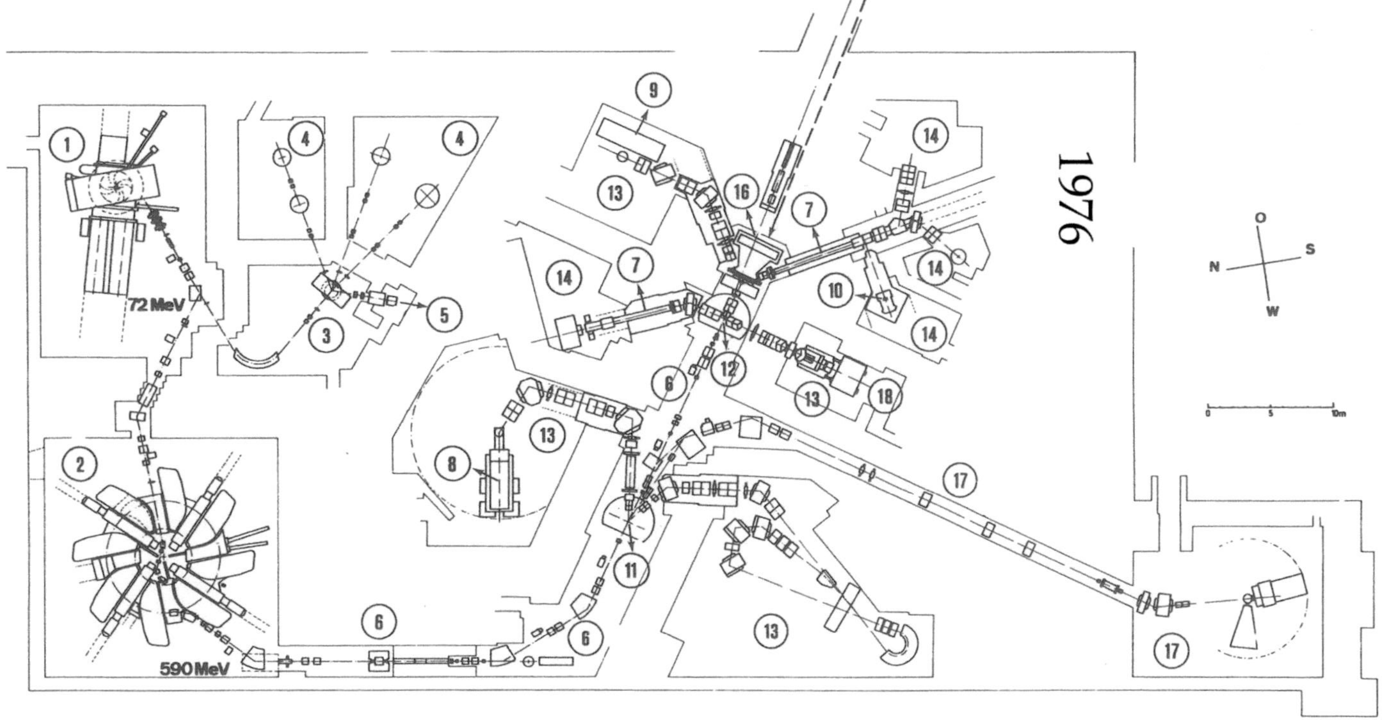

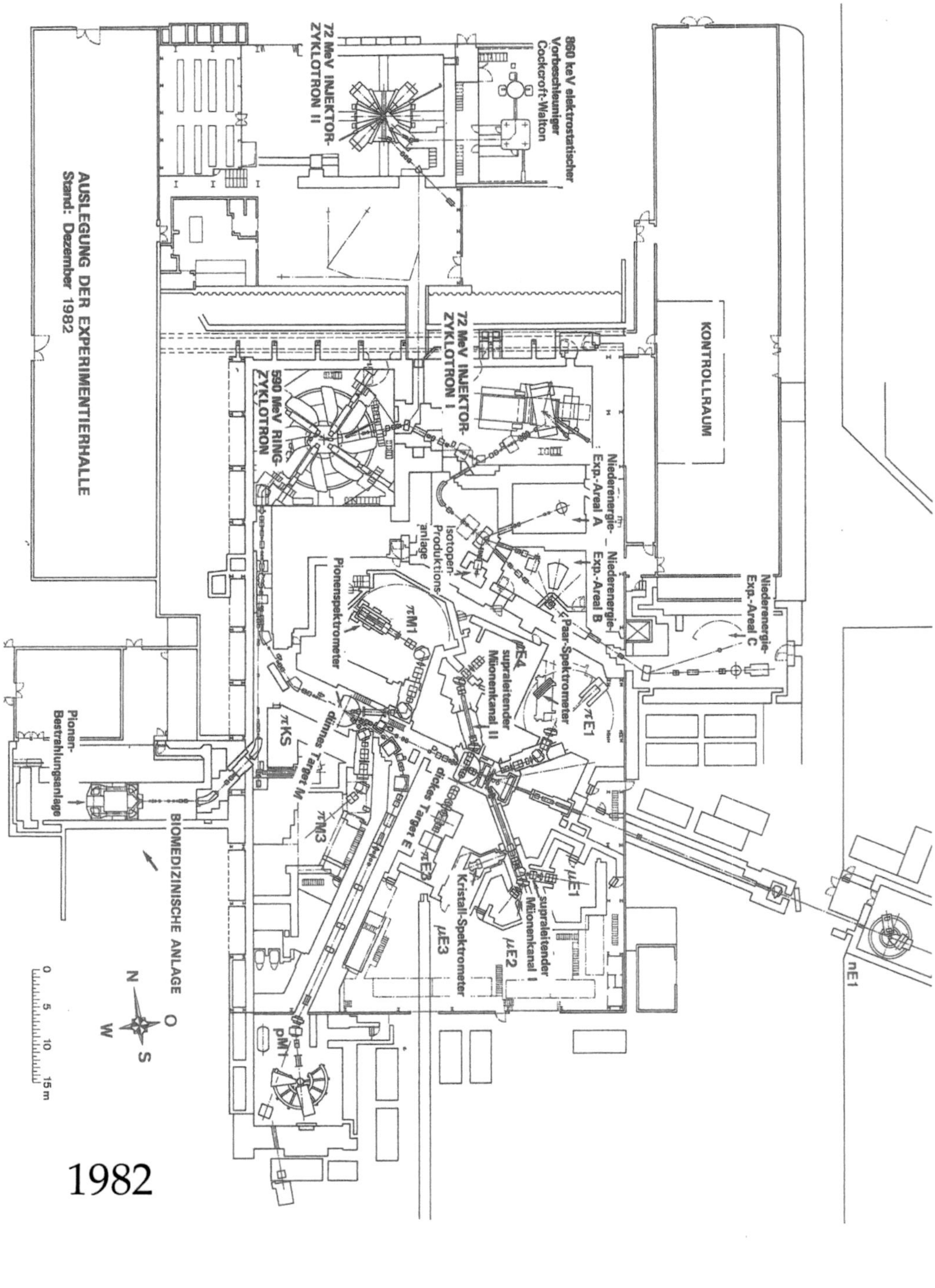
860 keV elektrostatischer Vorbeschleuniger Cockcroft-Walton
72 MeV INJEKTOR-ZYKLOTRON II
AUSLEGUNG DER EXPERIMENTIERHALLE
Stand: Dezember 1982
72 MeV INJEKTOR-ZYKLOTRON I
KONTROLLRAUM
590 MeV RING-ZYKLOTRON
Niederenergie-Exp.-Areal A
Niederenergie-Exp.-Areal B
Niederenergie-Exp.-Areal C
Isotopen-Produktions-anlage
Pionenspektrometer
πM1
Paar-Spektrometer
μE4
supraleitender Müonenkanal II
πE1
πKS
dünnes Target M
Pionen-Bestrahlungsanlage
dickes Target E
πM3
πE3
μE1
supraleitender Müonenkanal I
Kristall-Spektrometer
μE3
μE2
nE1
pM1
BIOMEDIZINISCHE ANLAGE
N O S W
0 5 10 15 m

1982

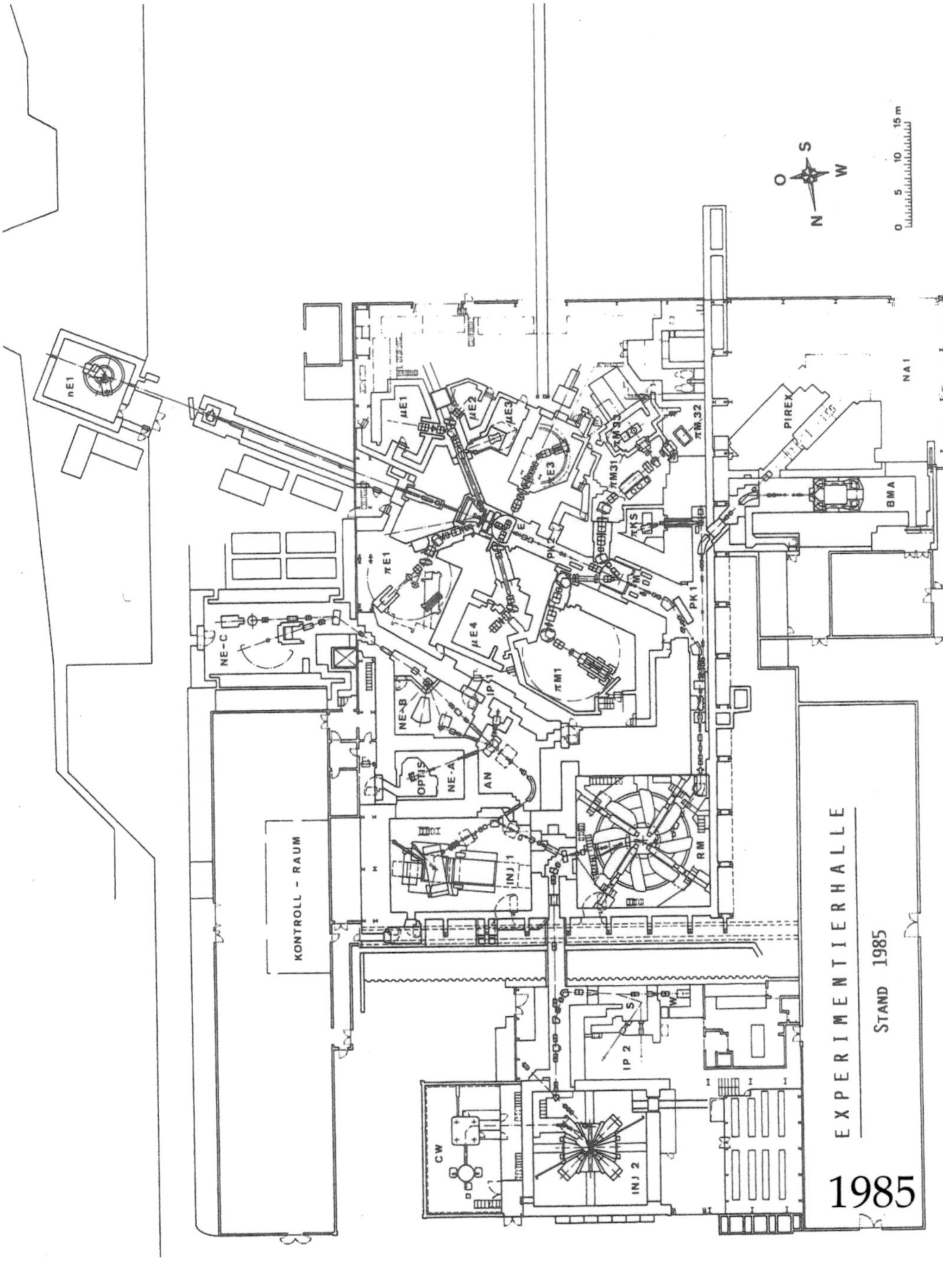
EXPERIMENTIERHALLE
STAND 1985
KONTROLL - RAUM
INJ 1
INJ 2
RM
CW
IP 2
NE-C
NE-A
NE-B
OPTIS
AN
IP 1
πM1
μE4
πE1
PK 1
PK 2
πKS
πM31
πM32
πM3
πE3
μE1
μE2
μE3
nE1
PIREX
NA1
BMA
O
S
W
N
0 5 10 15 m

1985

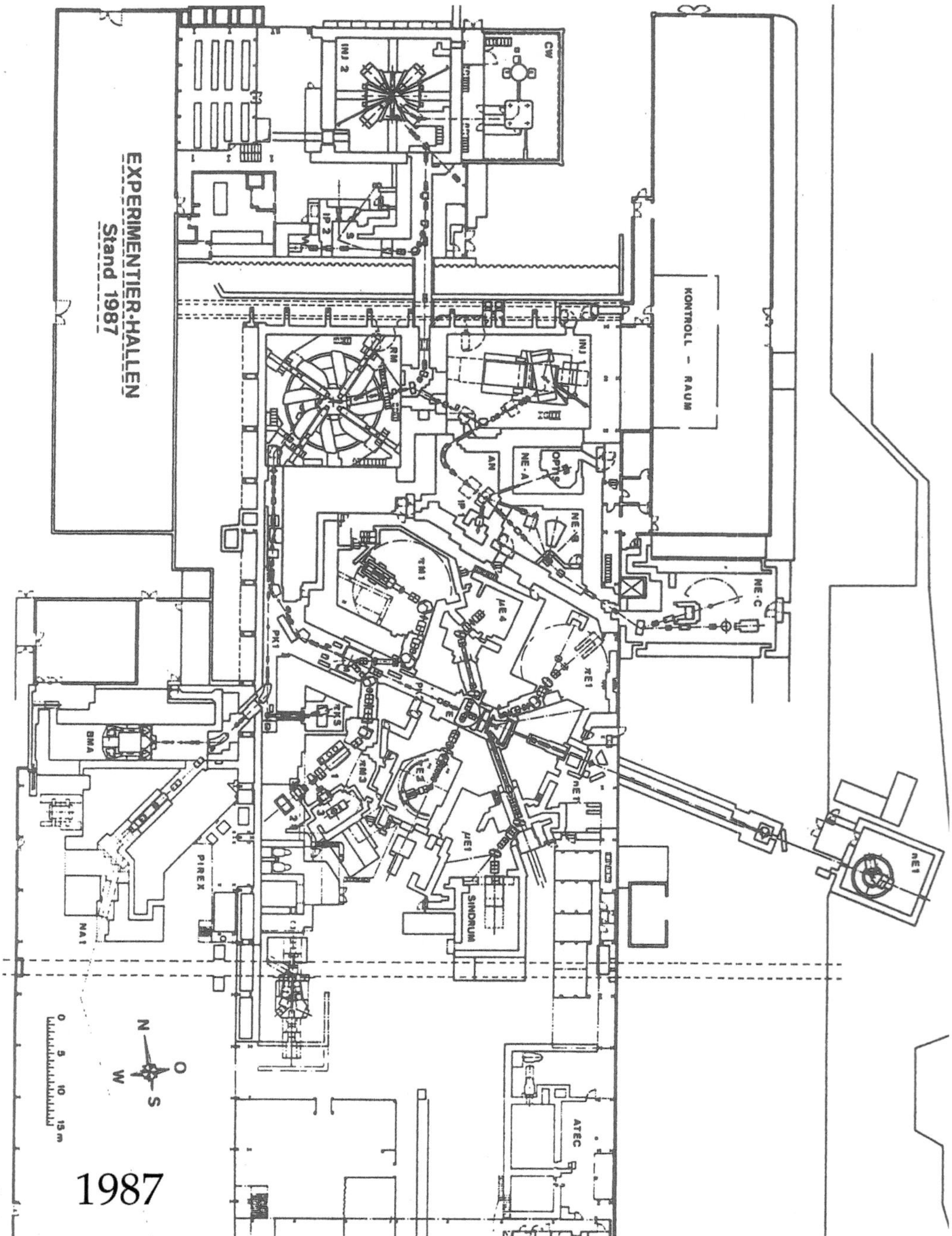
EXPERIMENTIER-HALLEN
Stand 1987
KONTROLL – RAUM
INJ 2
CW
IP 2
INJ 1
RM
AN
NE-A
NE-B
NE-C
PK1
SINDRUM
BMA
PIREX
NA1
ATEC
N
O
S
W
0
5
10
15 m
1987

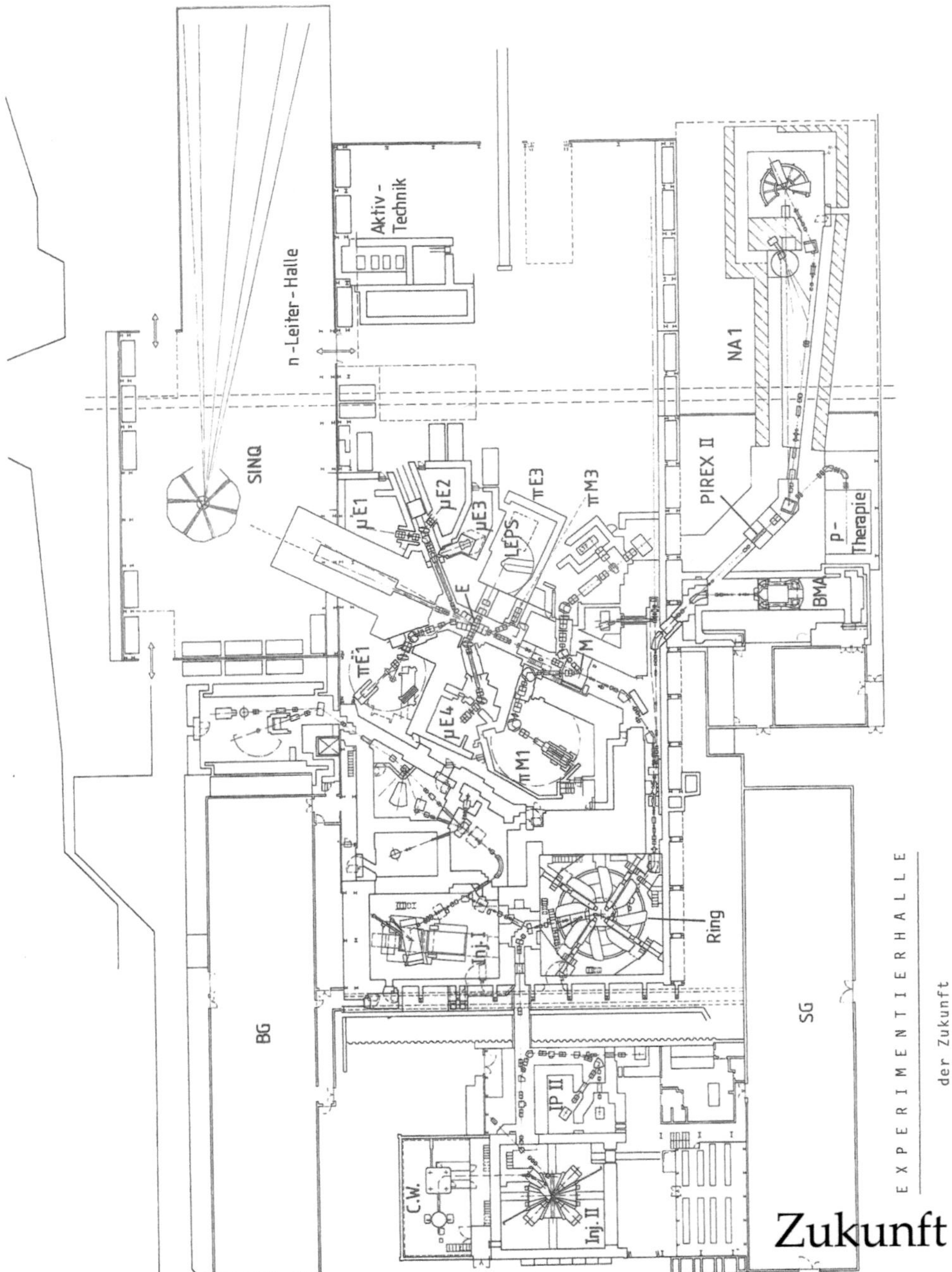

n-Leiter-Halle
Aktiv-Technik
SINQ
NA1
PIREX II
p-Therapie
BMA
µE1
µE2
µE3
LEPS
πE3
πM3
E
πE1
µE4
πM1
M
Ring
Inj. I
BG
SG
IP II
Inj. II
C.W.
EXPERIMENTIERHALLE
der Zukunft

Zukunft

Glossary of Technical Terms

Energy units in nuclear and particle physics

Electron volt (eV): Kinetic energy of a simply charged particle after it has been accelerated with a voltage of 1 volt. keV= kilo electron volt (acceleration voltage 1000 volts), MeV= mega electron volt (acceleration voltage 1 million volts), GeV= giga electron volt (acceleration voltage 1 billion volts), meV= milli electron volt (1/1000th of a volt).

Particles

Proton: Stable, positively charged elementary particle, nucleus of the hydrogen atom. Often used in particle accelerators as a projectile for the manufacture of other particles.

Neutron: Electrically neutral elementary particle. Along with the proton, it is part of the atomic nucleus. As long as a neutron is not bound in an atomic nucleus ("free" neutron") it is unstable.

Nucleon: Part of the atomic nucleus, either a proton or a neutron.

Electron: Negatively charged elementary particle. Atoms are made up of an atomic nucleus and a shell made up of electrons.

Positron: Positively charged electron. If electrons collide with positrons at high energy (about 5 GeV), they manufacture B mesons. Positrons are produced through certain radioactive decays and can be applied in medical diagnostics (positron emission tomography PET).

Mesons: Unstable elementary particles of a short life span, produced through high-energy particle collisions.

π-meson or pion: Is produced with the SIN accelerator by bombarding materials such as carbon or beryllium. Used in nuclear and particle physics, and also earlier in cancer therapies.

μ-meson or muon: Decay product of the pion. It is similar to the electron, but heavier. In addition to applications in nuclear and particle physics, it is used in particular as a probe in solid-state physics and chemistry.

Deuteron: Heavy hydrogen nucleus made up of one proton and one neutron.

Alpha particle: Helium-4 nucleus made up of two protons and two neutrons.

Polarized particle beams: The particle spin is uniform.

Particle Accelerators

Particle accelerators are machines in which charged particles are accelerated to high speeds through electric or electro-magnetic fields. To achieve this, there has to be a vacuum inside the machine.

Cockroft-Walton Generator: Facility for the production of high direct-current voltage. It works by using a so-called cascade connection which was developed by Heinrich Greinacher.

Van-de-Graaff Generator, also called band generator: Facility for the production of high direct-current voltage, invented in 1929 by Robert van de Graaff. It is often used as a particle accelerator at university institutes.

Linear accelerator or Linac: Accelerates particles along a straight line through successive drift tubes (electrodes) using alternating voltage. The principle goes back to Rolf Wideröe.

Cyclotron: Circular particle accelerator. Unlike in a linear accelerator, the particles are brought into a spiral-like path with magnetic fields so that they pass through one or several acceleration paths more than once. It was invented by Ernest Orlando Lawrence.

Synchrotron: Circular particle accelerator. An appropriately synchronized high-frequency alternating magnetic field (microwaves) are used. The particles – depending on energy level reached – are guided through re-adjusted magnetic fields into a closed path and reach speeds that are close to the speed of light. Edwin McMillan of the USA and Vladmir Weksler of the Soviet Union independently developed the synchrotron.

Synchrocyclotron: Special cyclotron by Edwin McMillan in which the frequency of the high-frequency field is adjusted to the speed of the particles in order to compensate for the relative increase in particle mass, while in the classic cyclotron the frequency remains constant.

Radioactivity

Radioactive decay or nuclear decay is a property of unstable atomic nuclei which change spontaneously after release of energy. The energy is almost always released as ionizing rays, i.e., energy-rich particles and/or gamma radiation.

Neutron Source

It delivers neutrons for materials research where neutrons, as they pass through the material, "see" only atomic nuclei. Neutrons are made by nuclear fission (in a reactor) or by spallation (in a spallation source) and have to be slowed (moderated) in such a way that their speed, meaning their "wavelength", is appropriate for the material to be examined. Prevalent are thermal neutrons with an energy under 100 meV whose speed is equal to a temperature of about 300 degrees Kelvin (room temperature). If neutrons are slowed even further (cold or ultra-cold neutrons), their energy decreases and their wavelength increases.

Synchrotron Light Source

Synchrotron beams – very intensive beams with light wavelengths between ultraviolet and x-rays – are created when electrons, as in a synchrotron, are forced into a circular path by means of magnetic fields. These beams can be applied to materials sciences and biology.

Sources and Photo Credits

Science History

H. Wäffler: "Forschung in Kernphysik an der ETH Zürich – 1928 to 1960", Paul Scherrer Institute, Villigen 1991 [nuclear physics at the ETH Zurich, Cyclotron Planning Team]

K. Alder: "Gedenkveranstaltung zum 100. Geburtstag von Paul Scherrer", Paul Scherrer Institute, Villigen 1990 [nuclear physics at the ETH Zurich, Cyclotron Planning Team]

A. Hermann, L. Belloni, U. Mersits, D.Pestre, J. Krige: "History of CERN Vol I", North Holland, 1987 [Beginnings of CERN]

Ch P. Enz, B. Glaus, G. Oberkofler: "Wolfgang Pauli aund sein Wirken an der ETH Zürich", vdf-Hochschulverlag an der ETH Zürich, 1997 [analysis of the ETH Board minutes in connection with high-energy physics and the selection of Jean-Pierre Blaser to the ETH]

P. Truöl: "Müonen und Pionen in Teilchenphysik und Anwendungen", Naturforschende Gesellschaft in Zürich, 2007 [SIN research programs]

P. Allenspach, R. Eichler, A. Furrer, K. Gabathuler, D. Herlach, W. Joho, E. Pedroni, A. Schenck, P. Schmelbach, Th. Stammbach, H.C. Walter: "25 Jahre Ringcyklotron am Paul Scherrer Institut – Geschichte, Ergebnisse, zukunft", SPECTRUM Spezial (Hauszeitung des Paul Scherrer Instituts), 1999 [SIN accelerator facility, research and applications]

Swiss Biographies

"Historisches Lexikon der Schweiz (HLS)", Stuftung HLS Foundation, since 1988; in electronic format www.hls-dhs-dss.ch

SIN Documents

SIN annual reports 1969 to 1987

SIN: "Zukünftige Aufgaben des Instituts, Stand der Arbeiten im Sommer 1970"

SIN :"Der neue Injektor für die SIN-Beschleunigeranlage", concept report, December 1973

SIN: "Das Schweizerische Institut für Nuklearforschung in Villigen - Ein nationales Forschungszentrum", an easy-to-understand brochure in German, published around 1974.

Legal Basis and Business of the Swiss Federal Councils

Federal newspaper ("Bundesblatt") as well as official bulletin of the federal assembly ("Amtliches Bulletin der Bundesversammlung"), in electronic format www.admin.ch/ch/d/ff/, part 1849 – June 1999 [building budget proposals for the ETH domain, federal science policy.]

Business of the Federal Council and the ETH Board

The documents and decisions cited in this manuscript are from personal archives. Their content, however, was communicated to the public.

Photo Credits

The numbers refer to the photo numbers.

AVES-Bulletin 25 (1988) : 61

Blaser Jean-Pierre : 39, 67, 69, cover photo front

CERN : 5

ETH-Zurich, Library, Picture Library : 1, 4

Joho Werner : 9, 11, 30, 35, 43, 44, 54, 75

Lawrence Berkeley National Laboratory : 2, 6

Markovits Christa : 32, 65

Olivo Miguel : 33, 66

PSI Picture Library : 10, 12, 13, 14, 15, 16, 17, 18, 19, 20, 23, 24, 25, 26, 27, 28, 29, 36, 40, 41, 42, 45, 46, 47, 51, 52, 53, 55, 60, 63, 64, 68

PSI Annual Report : 76

SIN Annual Report : 57, 59, 62, 70, 72, 73, 74

Steiner Erich : 21, 22, 31, 34, 37, 38, 48, 49, 50, 58

Steiner Lis: 56

Wikipedia : 3, 7, 8

Index of Persons

Not all first names could be determined from available documents.

A

B

C

D

E

F

G

H

I

J

K

S

T

U

V

W

Z

More books by Andreas Pritzker
are presented on
www.munda.ch